Design for Good Acoustics

J. E. MOORE, FRIBA

DESIGN FOR GOOD ACOUSTICS

WITH A FOREWORD BY
EDWARD D. MILLS, CBE, FRIBA, MSIA

ARCHITECTURAL PRESS: LONDON

Printed and bound by Staples Printers Limited, Rochester, Kent

FOREWORD

Most books on acoustics are written by scientists and deal with the subject from a scientific point of view. While this is obviously desirable in many cases, it does not always provide the architect with the material he needs when considering the design of a building in its early stages.

The publication of 'Design for Good Acoustics' is therefore doubly welcome, as it is not only written by an architect who understands acoustic problems, but has been prepared as a working tool to be used by architects and architectural students when they are considering the initial stages in the design of a building. The clear and concise way in which the information is presented, will commend it to all concerned with the design of buildings, and the fact that much of the information is presented in a visual form by means of plans, diagrams and drawings, makes it easy to use.

Good acoustics in a building are not merely a matter of applying some patent sound absorbent material to walls or ceilings, but are fundamental to the design of the building. Size, shape and volume are all-important factors which have a great bearing on the acoustics, and it is this basic approach to the problem which Mr. Moore's book stresses, and which makes it such an important publication.

I am sure it will be warmly welcomed by architects and architectural students, and will rapidly gain recognition as one of the standard working tools in relation to acoustics for all who are concerned with the design of good buildings.

EDWARD D. MILLS

CONTENTS

ACKNOWLEDGEMENTS

The author wishes to acknowledge his indebtedness to the authors of the books listed below and, for the reader who wishes to go more deeply into the subject, the list is offered as a bibliography. He would also like to thank Mr. Edward Mills for consenting to write a foreword, and Mr. William Apps for checking the original typescript.

THE ACOUSTICS OF BUILDINGS: A. H. Davis and G. W. C. Kaye. 1927

PLANNING FOR GOOD ACOUSTICS: H. Bagenal and A. Wood. 1931

ACOUSTICS AND ARCHITECTURE: P. E. Sabine. 1932

PRACTICAL ACOUSTICS FOR THE CONSTRUCTOR: C. W. Glover. 1933

MODERN ACOUSTICS: A. H. Davis. 1934

ACOUSTICS: P. L. Marks. 1940

ACOUSTICS OF BUILDINGS: F. R. Watson. 1941

PRACTICAL ACOUSTICS: H. Bagenal. 1942

ACOUSTICS FOR ARCHITECTS: E. G. Richardson. 1945

THE PRACTICAL APPLICATION OF ACOUSTIC PRINCIPLES: D. J. W. Cullum. 1949

ACOUSTICAL DESIGNING IN ARCHITECTURE: V. O. Knudsen and C. M. Harris. 1950

SOUND INSULATION AND ROOM ACOUSTICS: P. V. Bruel. 1951

ACOUSTICS IN MODERN BUILDING PRACTICE: F. Ingerslev. 1952

ACOUSTICS, NOISE AND BUILDING: P. H. Parkin and H. R. Humphreys. 1958

INTRODUCTION

This book is offered to the busy architect and the hard-pressed student. It is not in any sense intended as an alternative to the detailed studies of the subject now available to the specialist. As the title suggests, the writer is primarily concerned to provide, as concisely as possible, a guide to the architect in the early stages of design, and before he brings in the expert.

If the general shape of an auditorium is determined before acoustic factors are considered, no amount of 'acoustic treatment' will make up for what has been lost in the initial stages of design. The room may in fact never be satisfactory.

Under the heading 'Examples and their Individual Requirements' are included all those rooms where people will be attempting to listen to speech or music, whether this be a classroom or a concert hall. In all cases acoustic considerations are worthy of attention at the outset.

The book has therefore been arranged in four sections as follows:

1: The Properties of Sound.

2: The Behaviour of Sound in an Enclosed Space.

3: Design for Good Acoustics.

4: Examples and their Individual Requirements.

All sections have been made as concise as possible and diagrams have been used where they can take the place of text. Although knowledge of what is contained in the first two sections is desirable, the architect who reads only section 3, and then turns up in section 4 the example in which he is immediately interested, should not go far wrong in the first stages of design.

It is appreciated that any attempt to condense a subject of such scientific magnitude into a book which it is hoped the architect will have time to read, will be subject to criticism on the score of over-simplification. Bertrand Russell has said, in this connection, that one can be accurate, or one can be intelligible, but that it is difficult to be accurate and intelligible at the same time. It is hoped that this book will at least be intelligible, and that any inaccuracy will not seriously affect the early stages of design.

1: THE PROPERTIES OF SOUND

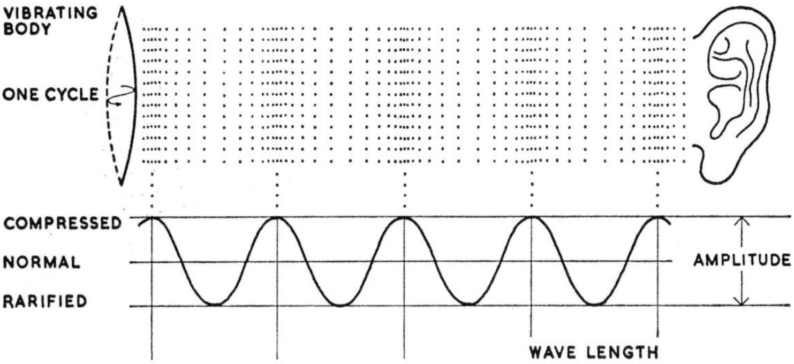

Fig 1

The nature of sound waves

Sound waves are alternate compressions and rarifications in the molecules of the air caused by a vibrating body.

Fig 1 illustrates this for a pure note. The outward movement of the vibrating membrane causes a compression in the adjoining molecules of air. When the membrane moves back the compressed air expands and causes a compression in the air adjoining the original centre of compression. Thus the compression travels outwards, followed by further compressions as the membrane continues to vibrate.

Spherical form of sound waves

All musical instruments, including the human voice, emit sound in all directions. The speed of sound being the same in all directions, undisturbed sound waves are spherical, as indicated in section in fig 2. The arrow drawn radially from the source of sound indicates the direction of outward movement of the sound waves, and is referred to as a 'sound ray'.

Sound waves reflected from a flat surface remain spherical, but are distorted by a shaped reflector, as will be seen later.

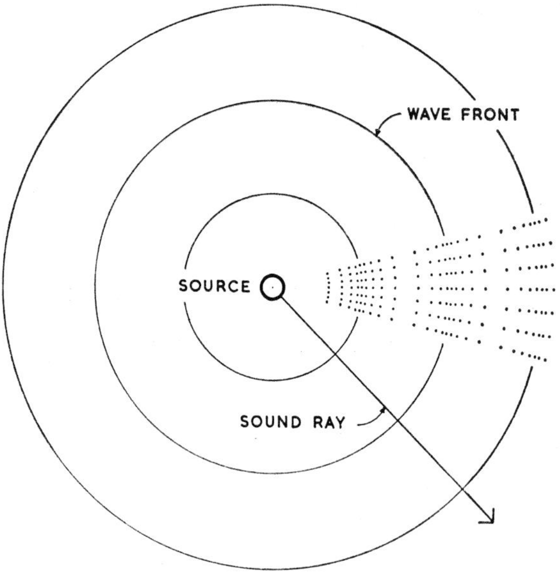
WAVE FRONT
SOURCE
SOUND RAY

Fig 2

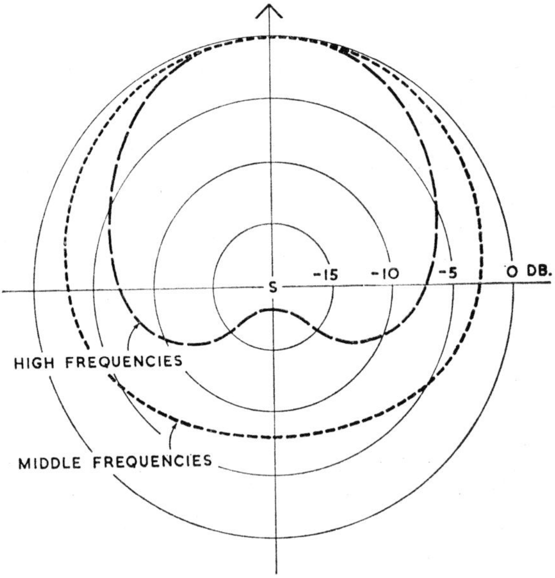
S
-15
-10
-5
0 DB.
HIGH FREQUENCIES
MIDDLE FREQUENCIES

Fig 3

Directional sound

Although musical instruments and the human voice emit sound in all directions, most are 'directional' in the sense that the sound waves emitted are strongest in one direction.

The graph, fig 3, indicates this in respect of the human voice. The speaker is facing in the direction of the arrow. It will be seen that at high frequencies there is a difference of 18 decibels between the sound emitted forwards and that emitted backwards, and a difference of 7 decibels in the sideways direction. This characteristic is important when considering the placing of reflectors.

Wavelength and frequency

The wavelength of a sound is the distance between the centres of compression of the sound waves, and is dependent upon the 'frequency' of the sound.

The frequency of a sound is the number of vibrations per second of the molecules of air, generated by the vibrating body. One complete movement to and fro of the vibrating body, and therefore of the molecules of air, is referred to as a 'cycle'. Thus frequency is expressed as the number of cycles per second.

High-frequency sounds have a short wavelength, and are heard as notes of high pitch. Conversely, low-frequency sounds have a long wavelength, and are heard as notes of low pitch. Some typical frequencies are listed below in round numbers.

MALE VOICE, vowel sounds,	about	100 c/s
MALE VOICE, sibilants,	,,	3000
BASS SINGER, bottom note,	,,	100
SOPRANO, top note,	,,	1200
PIANO, bottom note,	,,	25
PIANO, middle C,	,,	260
PIANO, top note,	,,	4200
PICCOLO, top note,	,,	4600
BASS VIOL, bottom note,	,,	40
ORCHESTRAL RANGE,	,,	45–4500
AUDIBLE RANGE,	,,	20–16000

Speed, wavelength and frequency

Sound waves travel in air at about 1100 feet per second, whatever their frequency or strength.

When an instrument vibrates, sound waves are generated at a frequency dependent upon the nature of the instrument. If we assume a sound of 100 c/s frequency and a listener 1100 feet away, then the first sound wave will reach the listener in one second, and the intervening space will be occupied by 100 following sound waves. The length of each wave will therefore be $\frac{1100}{100} = 11$ feet.

The wavelength of a sound can therefore be determined by dividing its frequency into the speed of sound. This will be seen to be of importance in the design of dispersive surfaces.

Loudness, sound pressure and sound level

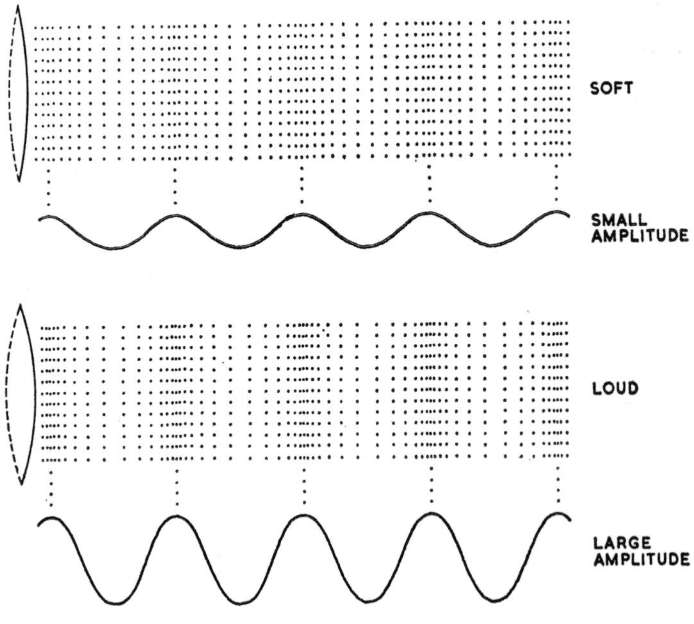

Fig 4

The greater the movement of a vibrating body, the greater will be the compression of the molecules of air, the greater the pressure on the ear drum, and the louder the sound. Fig 4 illustrates this and it will be noted that wavelength is not affected.

Sound, or air pressure is measured in dynes (a unit of weight) per sq. cm. For example:

Threshold of hearing: about 0·0003 dynes/sq. cm.
Threshold of pain: about 300 dynes/sq. cm.

It will be seen that the range of sounds which can be heard vary by a million times in their air pressure. To specify loudness in dynes/sq. cm. would thus involve unwieldy figures but, more important, such figures would not take into account the fact that the ear does not respond equally to changes of pressure at all levels of intensity. In other words, a very small increase in air pressure can be distinguished by the ear at low levels of intensity, but at high levels of intensity the increase must be considerable.

Loudness or 'sound level' is therefore measured on a logarithmic scale of 'decibels', the result of which is shown by the following figures:

Sound pressures in dyn./sq. cm.		Increase in sound pressure	Increase in sound level
First sound	Second sound		
0·01	0·0112	0·0012	1 db.
100·0	112·0	12·0	1 db.

The increase in loudness between the above pairs of sounds would seem to the ear to be about the same, in fact just perceptible in both cases. But the increase in *sound pressure* is far greater in the second case than in the first. In both cases the increase in *sound level* is one decibel.

At any level of intensity the sound pressure must be multiplied by 1·12 to make an increase of one decibel. An increase of 10 decibels sounds about twice as loud.

The following are examples of familiar sounds expressed in decibels:

	Decibels
THRESHOLD OF PAIN	130
Full orchestra, loud passage:	95
Loud speech, 3 feet distant:	70
STANDARD SOUND for comparing reverberation	60
Conversational speech, 3 feet distant:	50
Attentive theatre audience, total sound:	40
Faint whisper, 3 feet distant:	30
Rustle of one programme, 25 feet distant	15
THRESHOLD OF HEARING:	0

There are two other scales of measurement for sound level:

PHONS: a scale which takes into account the varying sensitivity of the ear to sounds of different frequencies and,

SONES: a scale which gives the proportional apparent loudness of sounds: for example, a sound of 10 sones would seem twice as loud as one of 5 sones.

Sound and distance

Intensity or sound pressure varies inversely as the square of the distance from the source, due to the attenuation of the wave front as it expands spherically. In fig 5a it will be seen that the area of the wave front has increased four times in doubling its distance from the source.

But, although the sound pressure is a quarter, the sound level

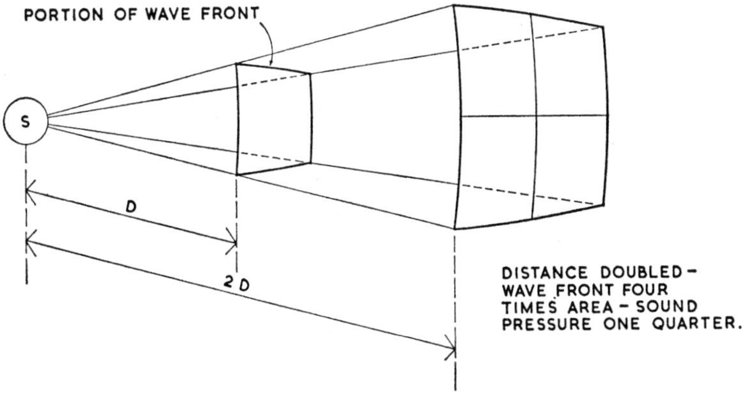

Fig 5a

will have dropped by only 6 decibels, and a drop of 10 decibels will be required for the sound to seem half as loud.

Nevertheless, because of the masking effect of background noise, (an average of 40 decibels in an auditorium) this falling off of sound pressure becomes critical for sounds of low intensity. In auditoria of more than about 50 feet in length reflectors are therefore indispensible. In smaller rooms the ceiling should be considered as a reflector and therefore should be of a suitable height, as discussed later.

The diagram, fig 5b, may explain the matter more clearly. It represents a speaker addressing an audience in the open air. The graph below the section shows the fall in sound level of his voice at various distances from him. Thus, if he speaks loudly, at 48 feet distance his voice will be heard as if he were speaking quietly at a distance of 3 feet. This is shown by the upper curve on the graph. If, however, he speaks in a conversational tone, the lower curve shows that his voice will be heard at 48 feet as a very faint whisper, and will be practically unintelligible at 96 feet.

All this assumes that there is no extraneous noise to interfere with hearing, that the speaker is always facing his audience, and that there is no obstruction between the speaker and the listener. The first of these conditions can never be satisfied as long as there is an audience present; the second condition will only occur in certain cases; the third condition is dependent on the rake of the seating. The situation can therefore generally be said to be not so good as that indicated by the graph, but will be improved if the auditorium is enclosed so that the audience hears reflections of the speaker's voice in addition to sound by direct path. If these reflecting surfaces are properly designed, a very considerable increase in sound level will be achieved.

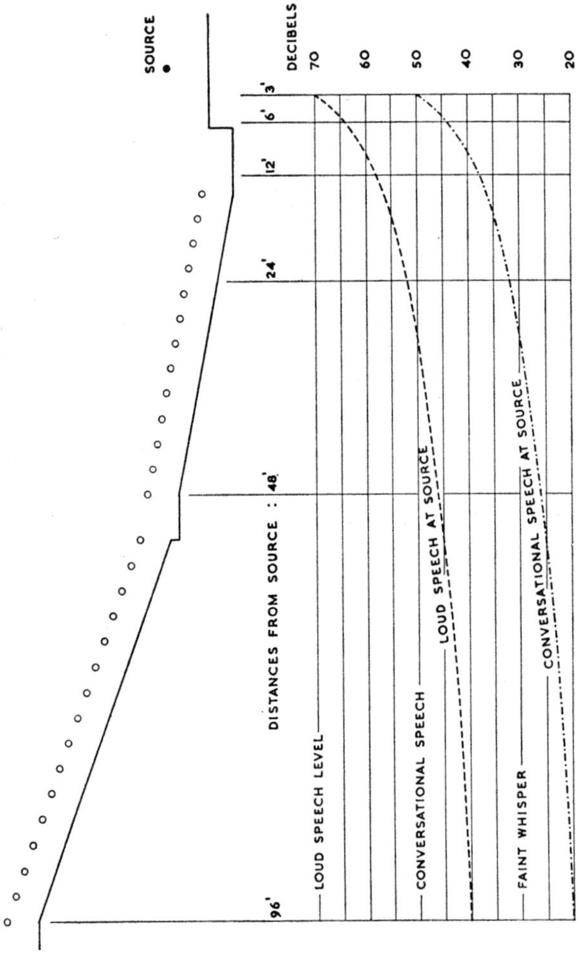

Fig 5b

2 : THE BEHAVIOUR OF SOUND IN AN ENCLOSED SPACE

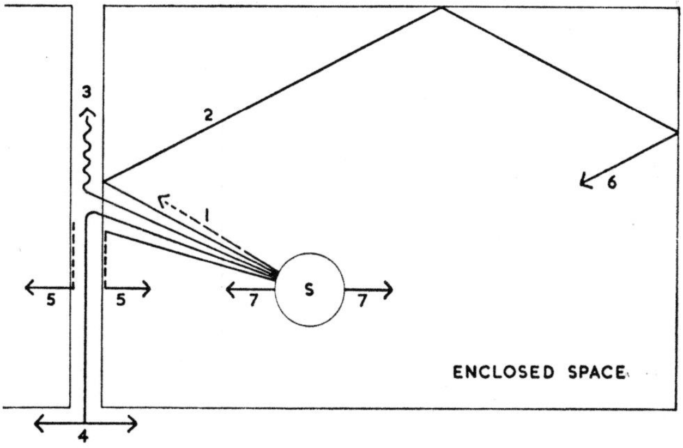

Fig 6

When sound is generated in a room it is reflected, absorbed and transmitted in various proportions in accordance with the nature of the construction. Fig 6 attempts to analyse the ways in which this occurs, and the following schedule is numbered to read with the drawing.

1: Sound absorbed in the air, also applicable to reflected sound.

2: Sound reflected from the wall surface.

3: Sound absorbed at the wall surface or its surface finish.

4: Sound conducted by the wall to other parts of the structure.

5: Sound emitted by resonance of the wall in both directions.

6: Sound inter-reflected between bounding surfaces, setting up reverberation.

7: Resonance of the enclosed volume of air by direct cross reflection.

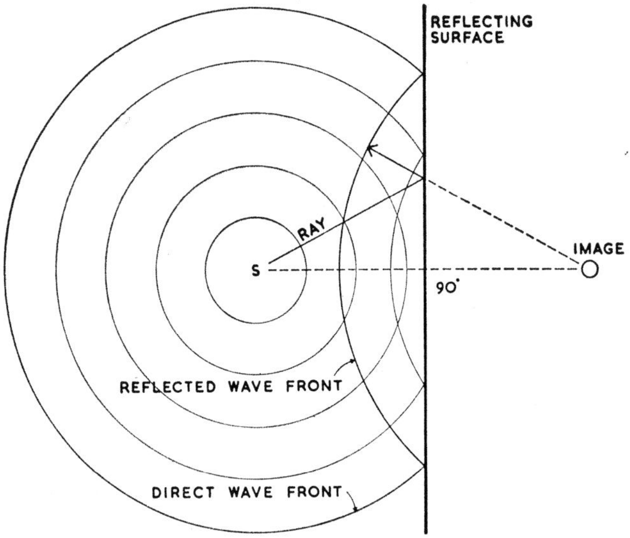

Fig 7

These aspects of the behaviour of sound, as they affect the acoustic design of auditoria, are dealt with in this section.

Reflections from a flat surface

Fig 7 illustrates the geometry of sound reflections from a flat surface. The reflected wave fronts are spherical and their centre of curvature is the 'image' of the source of sound. The image is on a line normal to the surface and at the same distance from the surface as the source.

Reflected sound rays

The first drawing in fig 8 shows that a reflected 'sound ray' is on a radial from the image in the case of a flat surface. The angle of reflection of the ray is equal to the angle of its incidence to the surface.

The second drawing shows that rays striking a curved surface

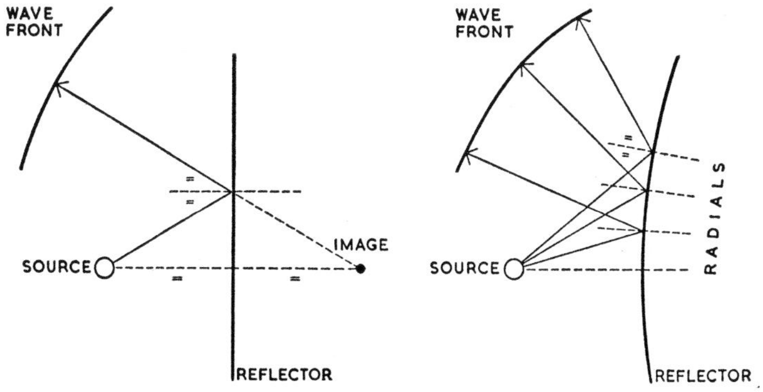

Fig 8

are each reflected so that the angle of reflection is equal to the angle of incidence to radials drawn at their points of contact. Each ray will in effect have its own image, the wave front will not be spherical, and must be obtained by drawing each ray of equal total length and joining their ends.

Reflections from curved surfaces

Figs 9, 10 and 11 make a direct comparison between the re-

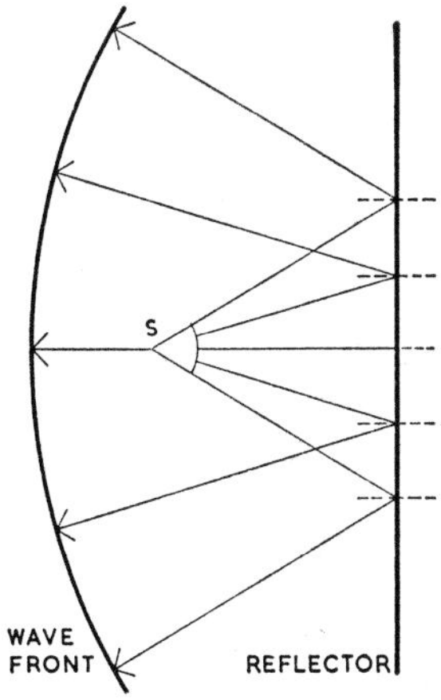

Fig 9

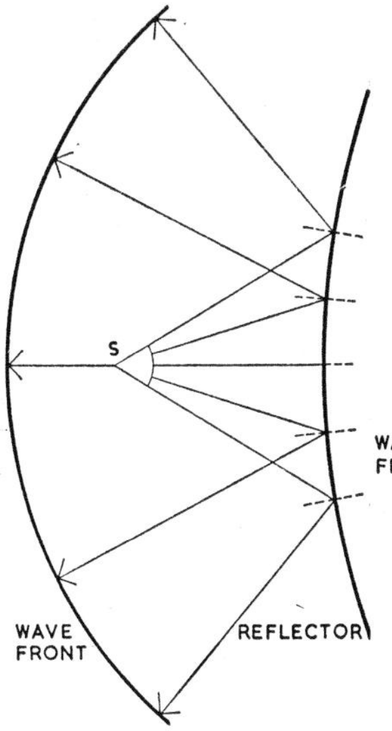

Fig 10

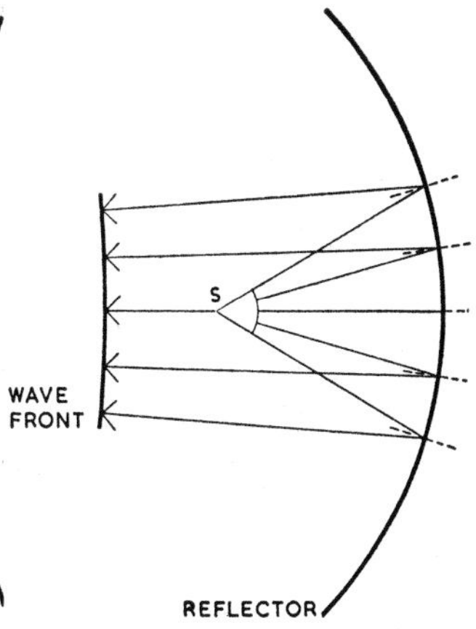

Fig 11

flections from flat, convex and concave surfaces. The distance from the source to the reflector is the same in each case, the cone of sound considered is the same, and the time interval at which the wave fronts are drawn is the same.

It will be seen, however, that the wave front from the convex surface is considerably bigger than that from the flat surface, and the wave front from the concave surface is considerably smaller.

It follows then that sound waves reflected from convex surfaces are more attenuated (and therefore weaker) and sound waves reflected from concave surfaces are more condensed (and therefore stronger) than is the case with a flat surface.

Sound reflected from a concave surface may eventually pass through a region of focus if the curvature of the surface is sufficient relative to the position of the source. Sound heard within this region may well be as loud as that heard close to the source. The effect of this will be seen in the next section.

Dispersion

Sound striking a modelled surface will be broken up into a number of small and weak waves, providing the modelling of the surface is bold enough. The first drawing in fig 12 shows

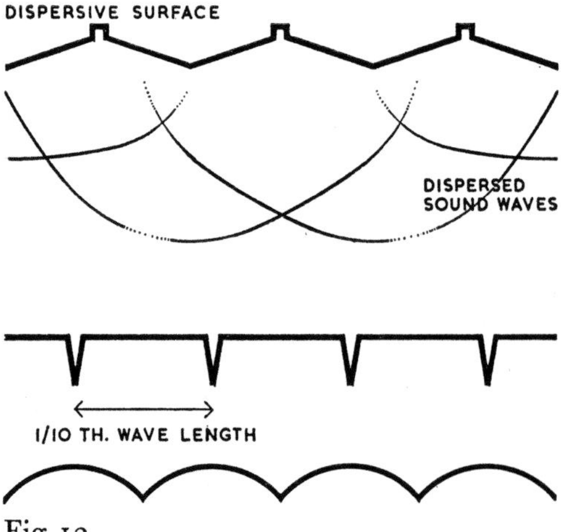

Fig 12

this, and it should be noted that the distance between the breaks in the surface must be at least a tenth of the wave length of the sound considered. Thus a surface with 6-inch breaks will disperse sound of 220 cycles per second and above, but will act like a flat reflector for sounds of lower frequency.

The scattering effect of dispersive surfaces can be employed to assist in the prevention of echoes or unwanted cross reflections.

Reflections from re-entrant angles

Sound entering a right-angled corner of a room will be reflected back towards the source, as shown at A in fig 13, if the adjacent surfaces are of reflective material.

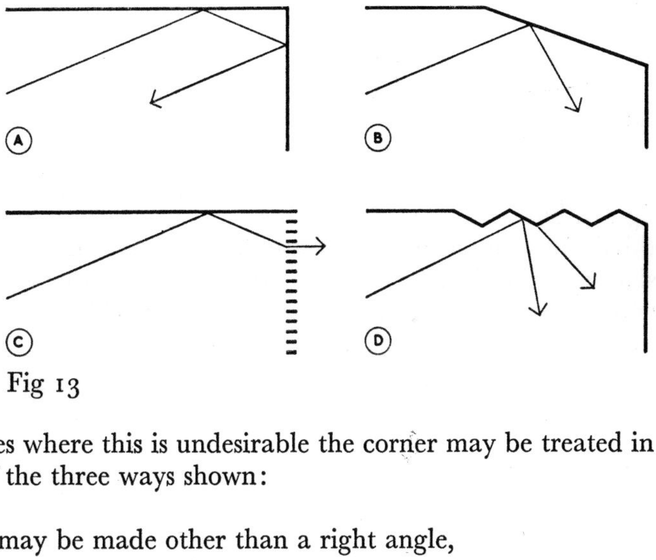

Fig 13

In cases where this is undesirable the corner may be treated in any of the three ways shown:

B: it may be made other than a right angle,

C: one surface may be made absorbent, or

D: one surface may be made dispersive.

Sound shadows

When sound is interrupted by an obstruction a sound 'shadow' is formed behind it, similar to a light shadow. Just as with light, however, diffraction occurs at the edge of the obstruction (E in

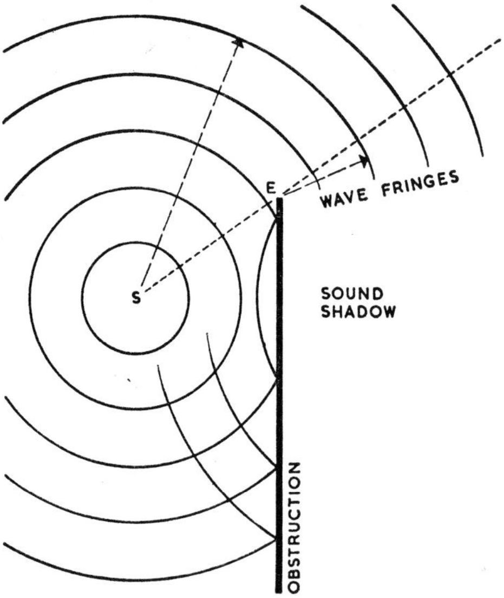

Fig 14

fig 14), but because of the much greater wavelengths of sound, this diffraction is considerable and wave 'fringes' are formed, as shown in the drawing. Sound shadows are, however, sufficiently well-defined to cause areas of poor audibility under the overhang of deep galleries.

Sound absorption

Sound generated in an auditorium is absorbed in four ways:

a: in the air,

b: at bounding surfaces,

c: in furnishings,

d: by the audience.

Air absorption

A small amount of sound is absorbed in the passage of direct and reflected sound through the air of a room. This is caused by the friction of the oscillating molecules of air and, although negligible at low frequencies, should be taken into account at frequencies above 1000 cycles per second when calculating reverberation periods.

Surface absorption

Absorption takes place whenever sound waves strike the bounding walls or surfaces of a room, and it occurs in a number of different ways, as follows:

a: by friction at the surface,

b: by 'penetration' in porous materials,

c: by molecular friction in resilient materials,

d: by molecular friction in a material during resonance,

e: by transmission 'through' the wall by resonance,

f: by conduction through the structure.

It follows that smooth, hard, dense and heavy materials absorb least sound, and rough, soft, porous and light materials absorb most.

The structure on which surface finishes are applied will naturally affect the total absorption occurring, as will be seen by reference to the factors listed above. This, however, is taken into account for average conditions when a material is tested, and the stated coefficient of absorption will include for this sufficiently for most practical purposes. In some cases the method of fixing and the backing material is stated.

A further type of absorption is provided by special acoustic materials based on the 'Helmoltz principle'. These are perforated or slotted materials backed by porous materials, such as wood or glass fibre. Absorption takes place by the resonance of the pocket of air in or behind each perforation.

Absorption by furnishings

Sound is also absorbed by furniture, curtains and any other such items which are present in the room. Coefficients of absorption are published for a limited range of furnishings, and manufacturers of theatre seats in some cases publish figures for the seat as a whole.

Absorption by the audience

The absorption of the audience itself is in most cases the largest single factor of absorption in a room, and is mainly due to the absorption of their clothing.

Because of this, room acoustics change perceptibly in accordance with the number of people present on various occasions. Since, however, each member of the audience is covering, and making inoperative, the absorption of a seat, a well upholstered seat will partly take his place acoustically, when he is absent. The introduction of highly absorbent seating will thus greatly reduce the variation in acoustic conditions due to changing numbers of audience.

The measurement of absorption

The unit of absorption still used in calculations is that devised by Professor Sabine towards the end of the nineteenth century. It is the amount of sound absorbed by an opening of one square foot, to which he gave the name 'open-window-unit' (later called a 'sabin'). The relation between percentage absorption, open-window-unit and coefficient of absorption is explained by the three examples shown in fig 15.

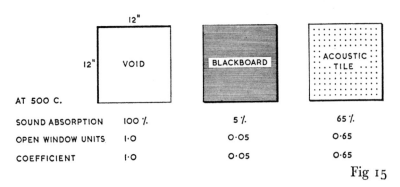

AT 500 C.			
SOUND ABSORPTION	100 %	5 %	65 %
OPEN WINDOW UNITS	1·0	0·05	0·65
COEFFICIENT	1·0	0·05	0·65

Fig 15

The total absorption of an auditorium may therefore be measured:

a: by multiplying the volume of air in the room by the coefficient of absorption per cubic foot,

b: by multiplying the total areas of all the various surfaces by their respective coefficients of absorption,

c: by multiplying the number of unoccupied seats by the coefficient of absorption for each seat,

d: by multiplying the number of people present in the room on any occasion by the average coefficient for each person,

and then, by totalling these amounts, the total absorption of the room can be stated in open-window-units.

Selective absorption

All materials absorb varying amounts of sound at different frequencies. Broadly speaking, porous and draped materials absorb preferentially at high frequencies, resilient materials at middle frequencies, and wood panelling at low frequencies. Materials designed on the Helmoltz principle generally absorb most sound in the middle or high frequency range, and in some cases absorption is very selective.

Since for all normal purposes the aim is to provide about the same degree of total absorption at all frequencies, it is necessary to 'mix' the materials used so that selective absorption is balanced out as much as possible in the end result.

Reverberation

Reverberation is the continuation of audible sound after the original sound has ceased, due to multiple reflections in an enclosed space.

It should not be confused with resonance, which is described later. Reverberation is a familiar phenomenon in a cathedral, where it may be up to ten seconds in duration for sounds of about 60 decibels. The resonance of wood panelling is, on the other hand, a matter of a fraction of a second.

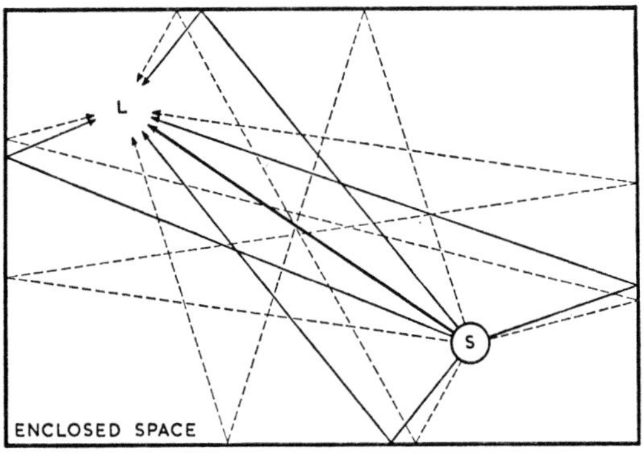

Fig 16

Fig 16 illustrates the initial stages of reverberation, only primary and secondary reflections being shown, but certain conclusions can be drawn from it. The listener L hears first of all, because by the shortest path, the direct sound from the source S, shown in a heavy solid line. He will next hear sound by four primary reflections from the wall surfaces, if they are plain and reflective, as well as reflections from the ceiling and possibly the floor. These reflections, shown by light solid lines, will arrive consecutively in accordance with the length of their travel. The same applies to the secondary reflections, shown by dotted lines, except that they will arrive later still. Inter-reflection between the bounding surfaces of the room will continue until the sound waves are so weak by attenuation and absorption that they become inaudible. Reverberation will then have ceased.

It follows therefore that, if the absorption of all surfaces is the same, the later the sound wave arrives at L the weaker it will be, and this is the reason for the gradual decay of reverberation in most cases. In cases where the absorption of the surfaces of a room varies considerably, or where very reflective surfaces are opposed to the main direction of the sound, reverberation will not decay steadily and may in fact be over-laid by stronger echoes or near-echoes.

Factors affecting reverberation

Generally speaking, however, it will be seen that the duration of reverberation depends upon:

a: the loudness of the original sound,

b: the absorbency of the bounding surfaces, furnishings and people,

c: the volume of the room and thus the length of the sound paths.

Loudness of sound tends to increase the period of reverberation, absorbent materials tend to reduce it, and greater volume tends towards an increase.

Sabine standard source of sound

For comparison purposes, however, a given loudness of sound is assumed, thus leaving the other two factors only to enter into the calculation and definition of reverberation. This 'standard sound' is still the one adopted by Professor Sabine in his first experiments, namely 60 decibels.

Thus 'time of reverberation' means the time taken for a sound of 60 decibels to decay to inaudibility.

Calculation of reverberation time

From the analysis of experimental results Sabine deduced the following formula for the calculation of reverberation time:

$$t = \frac{V}{A} \times 0{\cdot}05$$

where t is the time of reverberation in seconds, V is the volume of the room in cubic feet, and A is the total absorption of the room in open-window-units (or sabins). It will be seen that the disposition of the factors V and A in the formula agrees with what has been stated under 'Factors Affecting Reverberation' above.

This formula is sufficiently accurate for most cases. It becomes inaccurate:

a: when used for rooms with a very high proportion of absorbent material, and

b: when used in rooms of 'megaphone design' where nearly all sound is projected towards the absorbent audience and rear wall.

The more recent Eyring formula is an improvement as regards condition (a), but no formula can take into account the shape of the room. Apart from very 'dead' rooms, the simple Sabine formula will be sufficiently accurate for rooms designed correctly from other points of view.

Application of the Sabine formula

In most cases the best way to use the formula is as follows:

a: Design the room having all other acoustic considerations in mind, and in accordance with the requirements of the room generally.

b: Choose appropriate materials in order to prevent echoes, and to suit general architectural requirements.

c: Tabulate all materials, furnishings and occupants in order to arrive at a sum total of absorption for the room. This will be expressed in sabins, and an example table is given in Appendix A.

d: Invert the formula to read: $A=\frac{V}{t}\times 0{\cdot}05$ and find the required figure for A by inserting the volume of the room as designed, and the time of reverberation required.

e: Modify the materials employed in the room, or add absorbent materials, so that A, as designed, becomes A as required.

f: In some cases a modification of the volume may also be necessary.

It is usually sufficient if reverberation time is calculated at three frequencies: 125 c/s, 500 c/s and 2000 c/s. The aim is, in fact, to achieve an approximately constant period of reverberation over the whole musical range of frequencies.

Resonance

When sound waves strike the enclosing structure of a room it is set into vibration to a greater or less extent according to its nature. The materials vibrate at the same frequency as the incident sound waves, and in turn emit sound on both sides of the partition. This is how sound is heard 'through' a wall. Walls, partitions, floors, ceilings, panelling, etc., respond in this way – by sympathetic vibration or resonance. Heavy walls respond less than light partitions, and partitions less than panelling.

On the other hand, any given material will respond to a varying extent according to the frequency of the sound waves striking it. All materials have a 'dominant response frequency', the frequency at which they respond best, and if this is too sharply defined it may be disturbing acoustically.

Wood panelling responds over a range of frequencies in the lower register and, if the size of the panels is varied in a room, this frequency band can be broadened. In rooms for music, advantage is taken of this characteristic by the employment of extensive areas of wood panelling to give richness of tone. In concert halls it is an essential part of the design.

Because of resonance, music produced on a staged platform will be louder, as well as richer in tone, than if played on a platform of solid construction. A wood panelled apron to the platform will contribute also to the total resonance of the construction. Panelling around and near the source of sound will be more effective than that at a distance, being activated by stronger sound waves.

Finally it should be noted that resonance reinforces sound without appreciably prolonging it, as does reverberation. It may therefore be usefully employed in multi-purpose halls where a short period of reverberation is required for speech, yet some substitute for reverberation is desired in the case of music.

Air resonance

In rooms with parallel and plain opposite walls 'standing waves' may be set up by sounds of long wavelength. This occurs when the wavelength coincides with the distance between the walls, or is an exact fraction of this distance. A resonant effect is then produced which accentuates and prolongs sounds of certain frequencies, and creates a distortion in the balance of frequencies. Changes in pitch and vibrato effects may also occur during the decay of reverberation.

In large auditoria, with dimensions greater than about 35 feet, these effects will not be discernible. In smaller rooms, parallel and plain opposite walls should be avoided in design, especially where they flank the source of sound.

Sound transmission

Sound energy produced in a room is in part conducted through the structure to other parts of the building. A small proportion of sound is lost in this way and may be a nuisance to occupants of other rooms. Conversely, external noise can be transmitted to the interior of an auditorium by way of the structure or the foundations, thus reducing the value of good acoustic design.

Heavy structure transmits less sound than light construction. Transmission can also be checked by structural separation or isolation by absorbent materials.

Air-borne sound transmission may equally offset the value of good acoustic design and should be prevented by the provision of sound-absorbing lobbies, fixed, and preferably double, glazing. Sound-absorbing screens may often be employed effectively at points where extraneous sound may leak into an auditorium, but where sound lobbies would be inconvenient.

3 : DESIGN FOR GOOD ACOUSTICS

Best possible direct sound

Minimise the distance to rear seats by adopting square rather than oblong plan proportions, in so far as sight lines permit. The seating areas in figs 17, 18 are the same, but the distance to rear seats is reduced by a quarter in the second example.

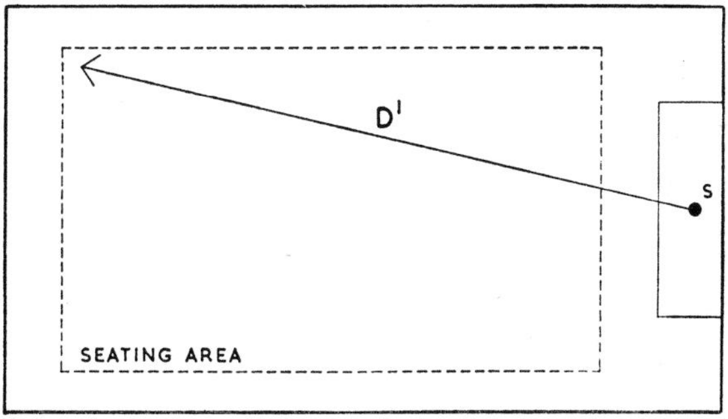

Fig 17

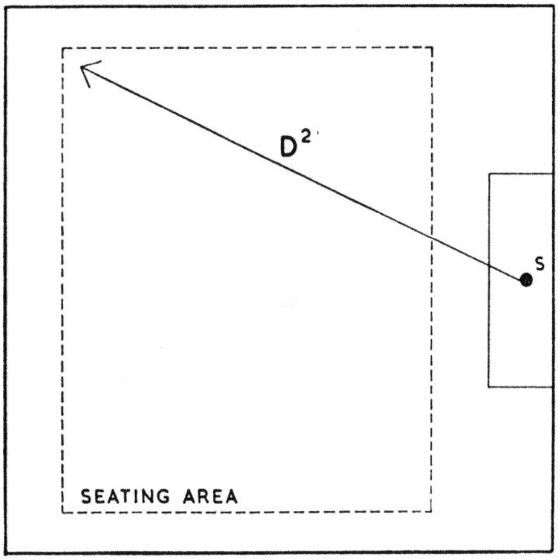

Fig 18

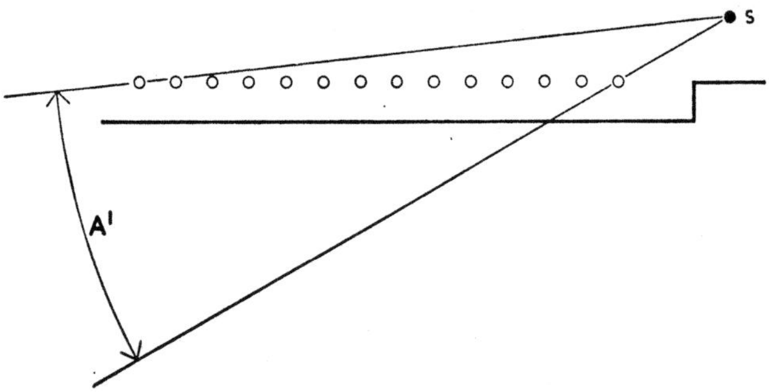

Fig 19

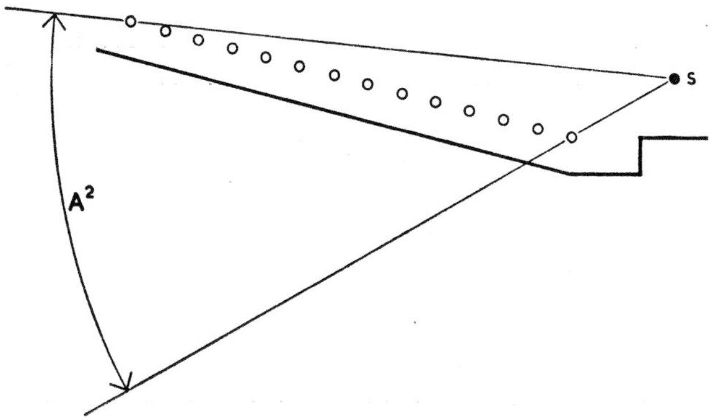

Fig 20

Economical arrangement of seats and gangways, and the introduction of balconies, also reduce the distance to rear seats, but sound shadows under balconies must be avoided, (see page 42).

Seating should be raked as much as possible, according to circumstances. The amount of sound shared by the audience in figs 19, 20 has been increased by a half in the second case.

Reinforcement by reflected sound

Where the source of sound is in one fixed position, as in figs 21, 22, the best reflector is near, above, and in advance of the source because:

a: the reflected sound paths are short and therefore reflections are strong and follow quickly upon the direct sound,

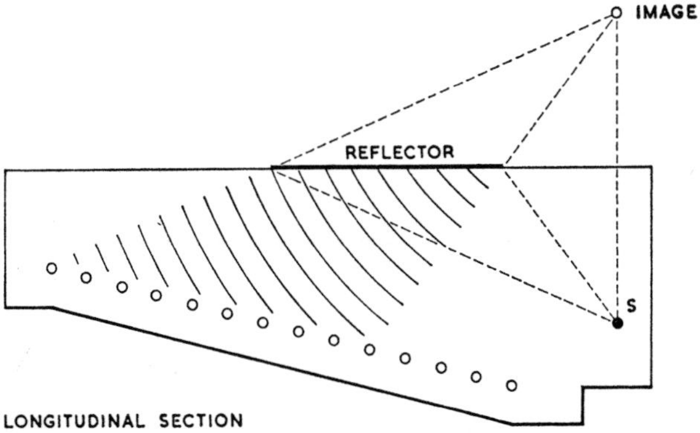

LONGITUDINAL SECTION

Fig 21

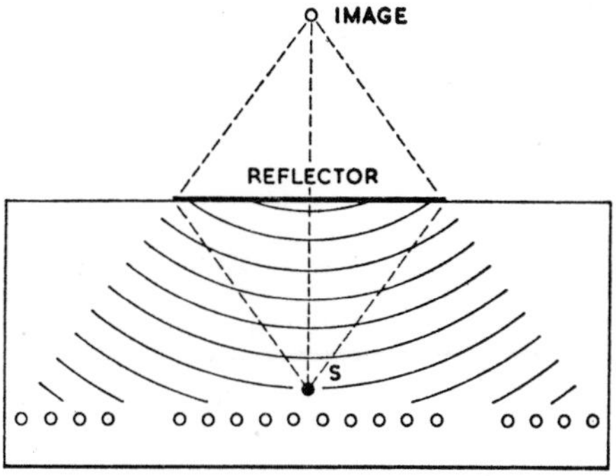

CROSS SECTION

Fig 22

b: unobstructed reflected sound benefits each member of the audience equally,

c: reflected sound is projected to the rear and middle rows of seats.

Where the position of the source is variable, but within a limited area, the reflector must be larger as indicated in figs 23, 24. In the longitudinal section, reflector AB is required for

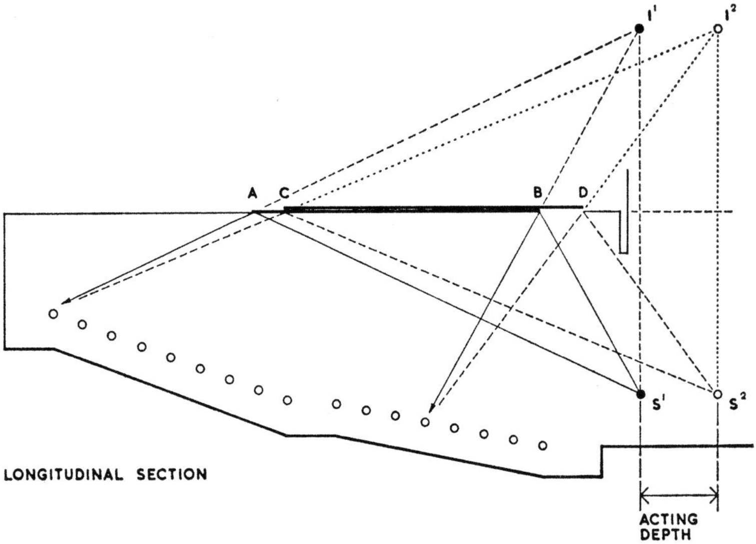

Fig 23

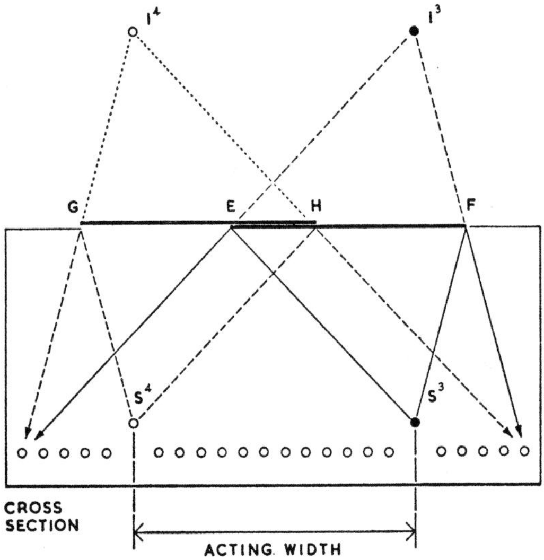

Fig 24

the forward position of the source, and CD for the rear position, if rear and middle rows of seats are to benefit. The length of the reflector should therefore be AD.

Similarly, the cross section shows that the width of the reflector

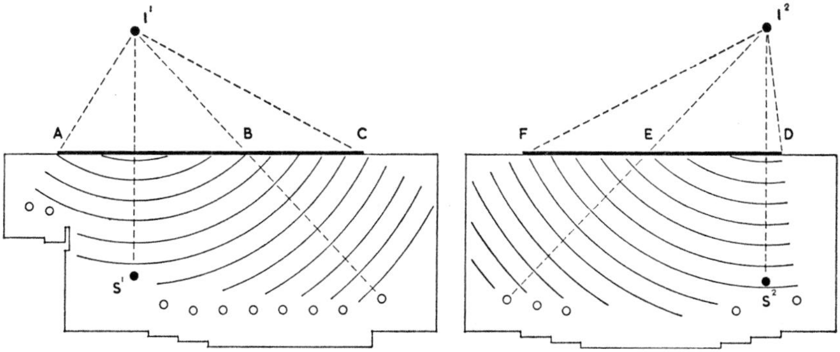

LONGITUDINAL SECTION CROSS SECTION

Fig 25

should be GF if the whole width of the audience is to receive reflections when the source moves laterally.

Where the source of sound may be anywhere in the room, as in a Council Chamber, the reflector should be near and above all positions of source, fig 25. In the longitudinal section, reflector AB will be sufficient when the source of sound is at S′. It must, however, be extended to C to allow for other positions of source. In the cross section, reflector FD will allow for a source of sound on either side of the room.

Where, as in 'theatre-in-the-round', the seating is on all sides of the stage, the actors will always have their backs turned to some part of the audience. If possible, reflectors should be arranged to project their voices back, as indicated in fig 26. The reflectors must, however, be within about 20 feet of the actors

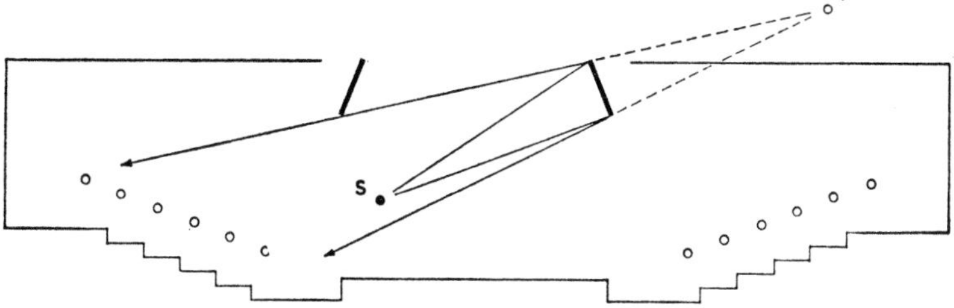

SECTION ON BOTH AXES

Fig 26

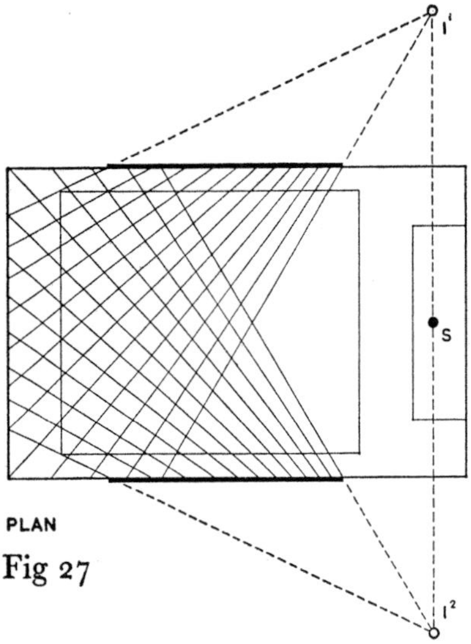

PLAN

Fig 27

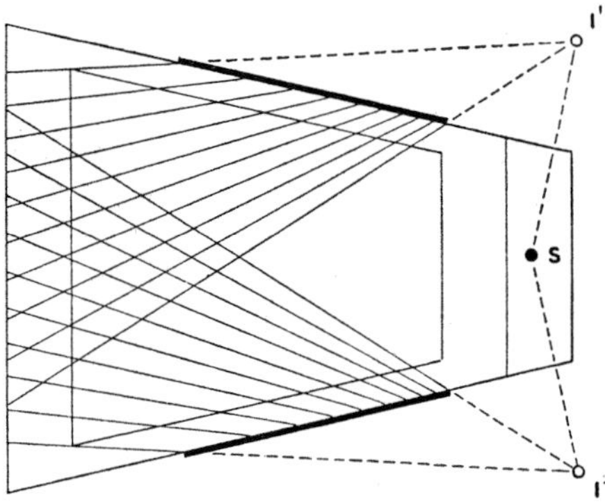

PLAN

Fig 28

to avoid echoes or 'near-echoes'. An alternative to this is to treat the room like a Council Chamber (fig 25) and provide a low horizontal reflector over the stage and first few rows of the audience. The size of the reflector would be worked out as in fig 25.

Where additional reinforcement of sound is required, the next best position after ceiling reflectors is at each side, and in advance, of the source, as shown in figs 27, 28. Such reflectors are less effective than those placed overhead because the reflections are received by the audience at a low angle of incidence. They are least effective when the audience is seated on a level, or nearly level, floor. Where the source of sound moves laterally to a considerable extent, as in a theatre, side wall reflectors may cause fluctuations in sound level as the source moves, due to the changing 'coverage' of the reflections. The effect of this should therefore be examined for extreme positions of source.

Controlled reinforcement by angled reflectors

Reflectors can be brought nearer the source, can be smaller, and can be arranged to project sound towards the rear rows of seats, by setting them at an angle to the source, as shown in

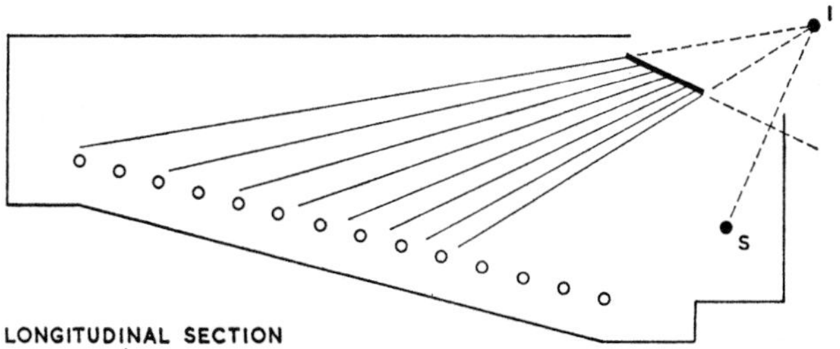

LONGITUDINAL SECTION

Fig 29

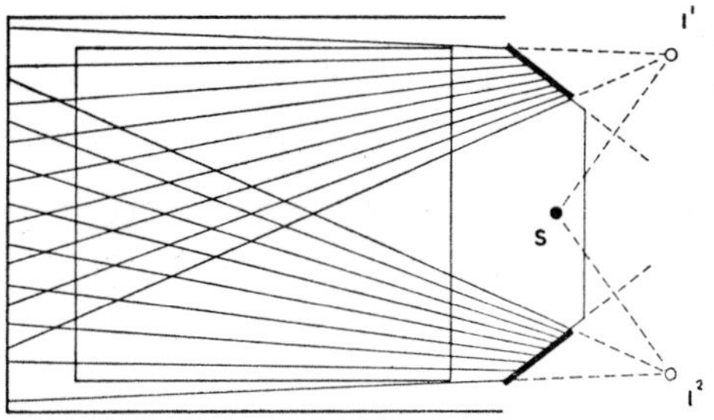

PLAN

Fig 30

figs 29, 30. The resulting reflections will follow more closely upon the direct sound, inasmuch as the reflector is nearer the source. As will be seen below, however, a ceiling reflector and an angled reflector can be combined in one design to further reinforce sound in rear seats.

Progressive reinforcement by multiple reflectors

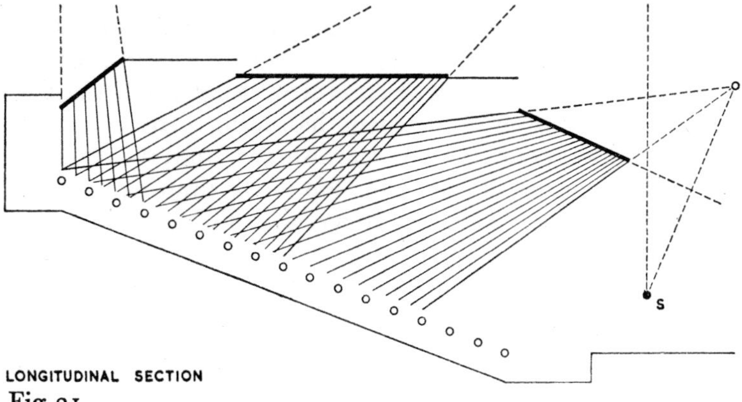

LONGITUDINAL SECTION

Fig 31

Direct sound can be reinforced progressively towards the rear of an auditorium by the use of multiple reflectors. An example is shown, fig 31, in which the rear four rows of seats receive three overlapping reflections, the next four rows two reflections, the next five rows one reflection, and the front four rows direct sound only.

Progressive reinforcement by shaped reflectors

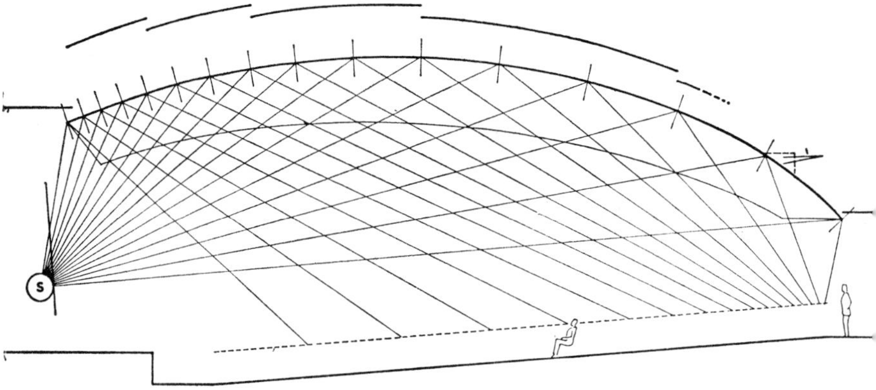

Fig 32

A similar progressive reinforcement of sound can be obtained by shaping the ceiling reflector so that the reflected sound waves are progressively stronger towards the rear of the auditorium.

The example shown in fig 32 would be appropriate to a cinema with the amplifier behind the screen. In the longitudinal section, sound rays are shown projected from the source at equal angles. After reflection they are, however, closer together at the rear than at the front of the room, indicating a strengthening of the sound waves towards the rear. Alternatively, the reflector may be 'broken', as shown at the top of the drawing, so that a proportion of the sound is dispersed. This would improve the quality of reverberation in the auditorium and allow for ventilation grilles and indirect lighting.

In the cross section, fig 33, the reflector is shaped so that the whole of the sound received by the reflector is projected evenly over the width of the seating.

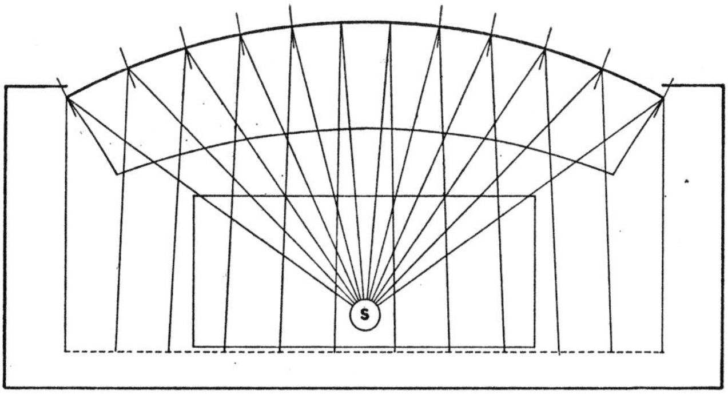

Fig 33

HAVING DETERMINED THE POSITION AND EXTENT OF THE REFLECTORS REQUIRED, ALL OTHER SURFACES SHOULD BE MADE DISPERSIVE OR ABSORBENT IN ACCORDANCE WITH WHAT FOLLOWS.

Prevention of echoes and near-echoes

Distinct echoes can occur in large auditoria due to delayed reflections. 'Near-echoes' are heard as an extension of the original sound, though the listener may only be conscious of a blurring of speech or music.

Figs 34, 35 illustrate an echo from the rear wall of a room, and a 'near-echo' from part of the side wall. In the first case the difference between direct and reflected sound path is about 80 feet, and in the second case about 40 feet.

Echoes can be distinguished when reflected sound is heard about one-fifteenth of a second or more after the direct sound, a difference in sound path of 70 feet or more.

'Near-echoes', sufficient to cause 'blurring', occur when this difference in sound path is between 35 and 70 feet. The blurring of speech will be understood by reference to the following time intervals for average speech:

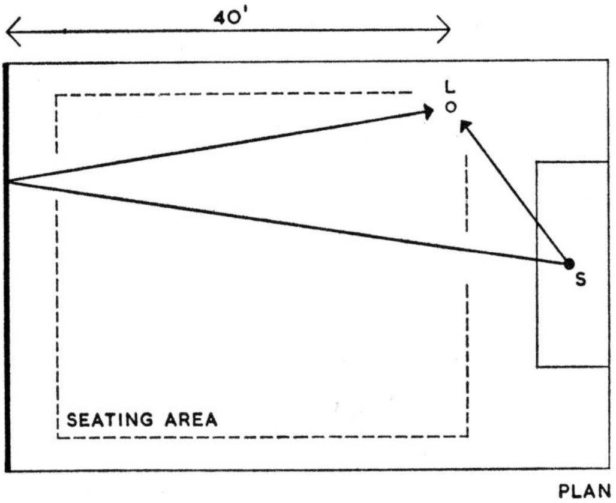

Fig 34

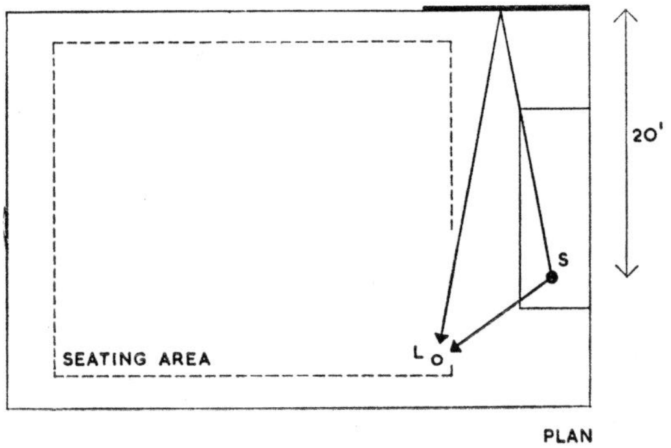

Fig 35

Duration of syllable: $\frac{1}{10}$ sec.

Gap between, if any: $\frac{1}{20}$ sec.

Gap between words: $\frac{1}{5}$ sec.

It will be seen that the time intervals in speech are of the same order as the time intervals of echoes and near-echoes. A similar loss of clarity in music is caused by 'near-echoes' and the effect of echoes needs no explanation.

Flat surfaces liable to cause echoes or 'near-echoes' should be made about 70 per cent absorbent or, if also dispersive, about 50 per cent absorbent.

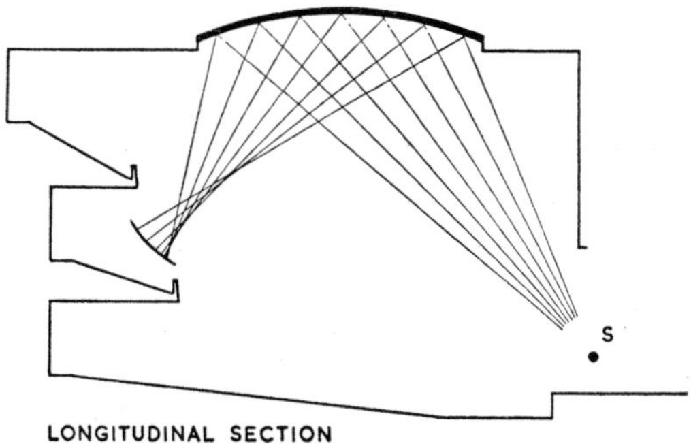

Fig 36

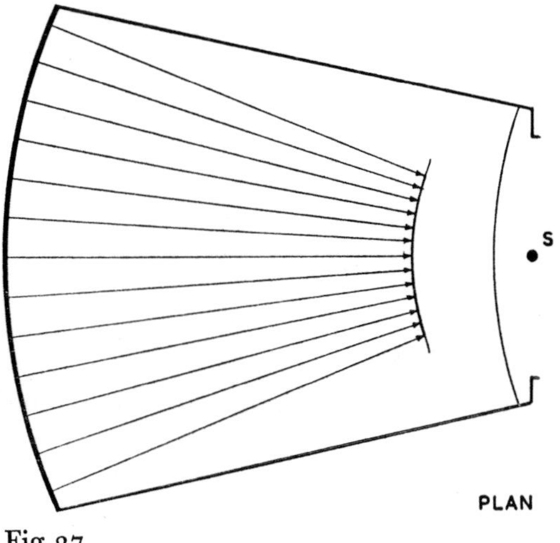

Fig 37

Concentrated echoes and reflections

Figs 36, 37 explain how echoes may be concentrated, and in some cases brought to a focus, by the introduction of large concave surfaces.

Such echoes may be as loud, or louder than, the original sound, and the surfaces producing them must be made at least 70 per cent absorbent *and* dispersive. Otherwise such shapes should not be employed.

It should also be noted that, since in most cases the source changes position, the number of seats affected by a concentrated echo will be greater than is indicated by analysis for a central position of source.

Reflectors should never be more than slightly curved in cross section. Even if the distance from the source is insufficient to cause an echo or near-echo, the result will be an unbalanced reinforcement of sound over the audience. Moreover, as the position of source changes, so will the loudness of sound change for any one member of the audience. Fig 38 illustrates a case where the curvature of the ceiling is too great.

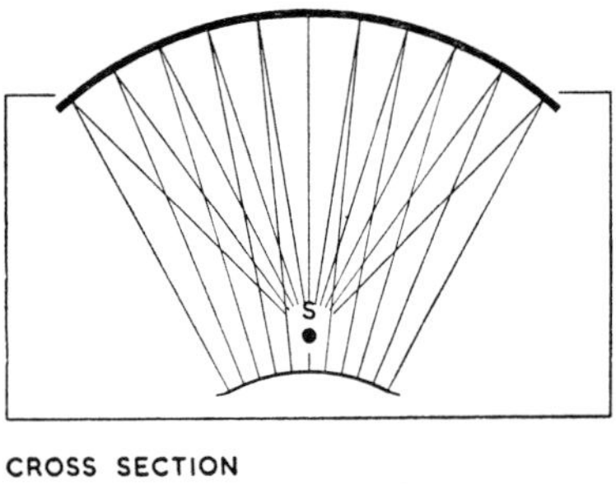

Fig 38

Corner echoes

Echoes may be produced by the introduction, at a distance from the source, of right-angled corners with both surfaces reflective. Figs 39, 40 illustrate two such cases. Diagrams B, C and D on page 23 show how such echoes can be avoided.

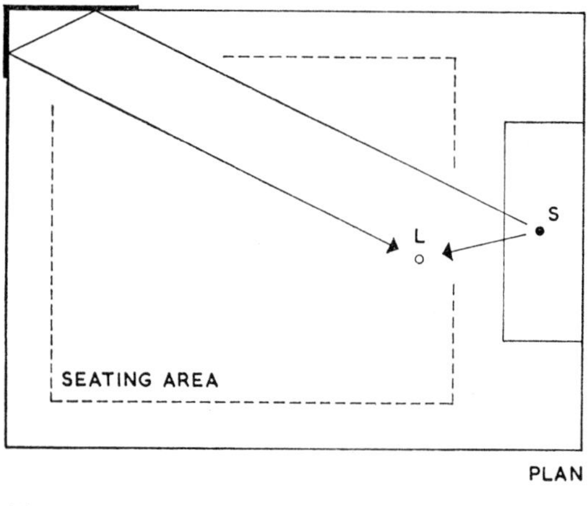

Fig 39

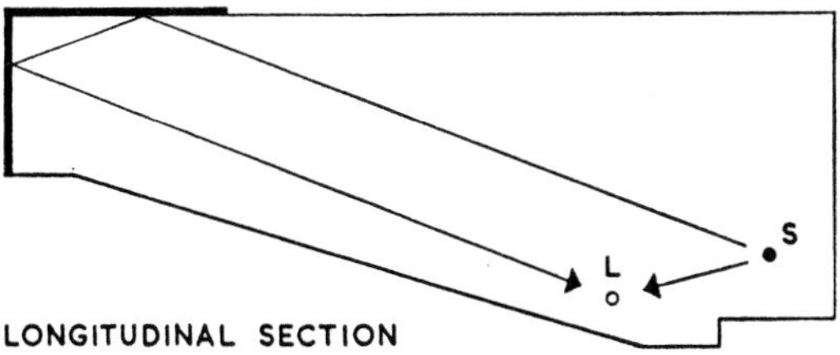

Fig 40

ROOMS MUST BE CAREFULLY CHECKED FOR SURFACES LIABLE TO CAUSE ECHOES OR NEAR-ECHOES, AND THIS MUST BE DONE FOR ALL POSSIBLE POSITIONS OF SOURCE.

Sound shadows

Sound shadows usually occur where balconies are too deep, preventing rear seats underneath from receiving much-needed ceiling reflections.

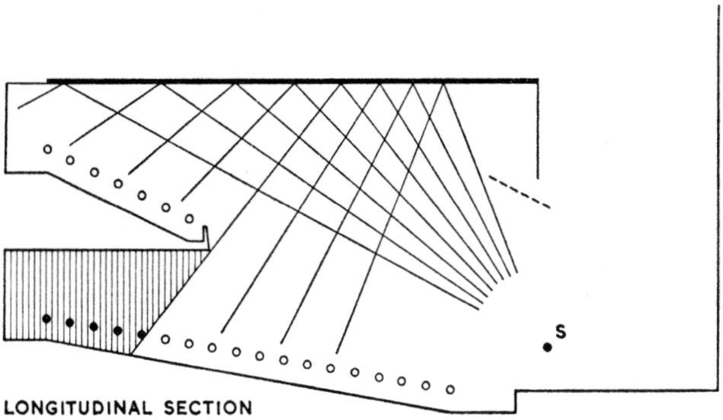

Fig 41

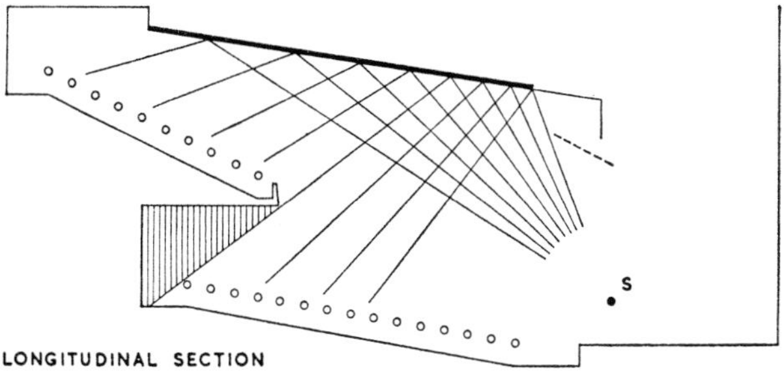

Fig 42

Fig 41 illustrates this, though the introduction of the angled reflector, shown dotted, would compensate for the lack of ceiling reflections.

An alternative is shown in fig 42, where seats have been removed from under the gallery and put at the rear of the gallery above. Provided the farthest seats are not, in the circumstances, too far from the stage, this will be more satisfactory.

Correct period of reverberation

Reverberation gives tonal quality to sound, but must not be excessive. Excessive reverberation reduces the clarity of music and speech by filling in the gaps between notes and syllables. Where reverberation is long, the earlier part is strong enough to cause a merging of consecutive sounds and has the same 'blurring' effect as near-echoes, described above. However, more reverberation is acceptable for music than for speech, and more for organ and choral music than for orchestral music.

Appropriate periods of reverberation are given below. Since a longer period of reverberation is expected by the listener in a large room than in a small one, two figures are given in each case. They refer to the range of volumes usually met with under each heading. Reverberation should be approximately the same at all frequencies.

SPEECH ONLY	$\frac{3}{4}$–1 sec.	COUNCIL CHAMBERS, LAW COURTS, LECTURE THEATRES, DEBATING HALLS.
TRAINED SPEAKERS & INCIDENTAL MUSIC	1–$1\frac{1}{4}$ sec.	THEATRES: PLAYS, MUSICAL COMEDY, AND VARIETY.
REPRODUCED SOUND	1 sec.	CINEMA, (*Reverberation added to sound track.*)
SOLO INSTRUMENTS & SMALL GROUPS	$1\frac{1}{4}$–$1\frac{1}{2}$ sec.	SMALL CONCERT HALLS.

MULTI-PURPOSE HALLS	$1\frac{1}{4}$ sec.	SCHOOL HALLS, COMMUNITY HALLS.
ORCHESTRAL MUSIC	$1\frac{1}{2}$–$2\frac{1}{4}$ sec.	LARGE CONCERT HALLS.
OPERA	$1\frac{1}{4}$–$1\frac{1}{2}$ sec.	OPERA HOUSES.
ORGAN AND CHOIR	$2\frac{1}{2}$–4 sec.	CHURCHES, CATHEDRALS, (*Chancel only.*)

Quality of reverberation

Reverberation should provide a background tone and should decay evenly. To effect this:

a: ensure strong direct sound and closely following primary reflections,

b: avoid opposite and parallel plain surfaces which can set up strong inter-reflections,

c: surfaces not used as reflectors should be either absorbent or dispersive, or both.

d: absorbent and dispersive surfaces should be well distributed.

Allowance for variable audience

The audience being the biggest factor of absorption in most cases, variations in its size will affect the length of reverberation. Wherever possible, therefore, use well-upholstered, absorbent seats. When empty, they will then partially compensate for absent members of the audience.

Apart from this, the auditorium should be designed to provide the ideal period of reverberation for what is expected to be the more general attendance, or what is deemed to be the more important conditions.

In this connection it should be borne in mind that in some cases the minimum attendance may be very small, yet acoustic conditions must still be reasonable. In theatres, for example, the minimum attendance is at rehearsals, and in the council chamber, a quorum.

Adjustable panels, absorbent on one side and reflective on the other, may be employed in order to allow for variations in the size of the audience, or to provide suitable conditions for the various uses of a multi-purpose hall. These will, however, cost money and their value will be dependent upon intelligent use. Reducing the size of an auditorium by curtains or screens will also reduce the reverberation time for small audiences.

Adequate resonance

Resonance gives richness of tone to music and speech. In certain instances, resonance reinforces sound; for example, a staged platform emits resonant sound which would otherwise be conducted away by one of more solid construction.

While all auditoria benefit from the use of suitable resonant material, in rooms for music the use of wood panelling is essential. The value of the resonant material is in the following order of effectiveness:

a: staged platform with panelled apron,

b: panelling around and in contact with the platform,

c: side wall panelling near the platform.

Wood panelling of various thicknesses and sizes responds over a wide range of frequencies. Materials which respond at one frequency only are quite unsuitable.

Standing waves

'Standing waves' are referred to under 'Air Resonance' on page 32. They may occur in the following situations where plain reflective surfaces are set opposite and parallel to each other:

a: between platform and horizontal ceiling above,

b: between the parallel sides of a platform recess,

c: between opposite sides of small rooms with plain walls.

Dispersive or absorbent treatment of one side will effectively prevent the trouble.

Materials*

In the design of an auditorium few of the finishes need be chosen solely for their acoustic properties. The aim should rather be to choose materials which are desirable from other points of view, both functional and aesthetic, and employ them appropriately for acoustic purposes. Plasters, panelling and curtains all play their part in an acoustic design, though they are not thought of as acoustic materials.

Where special acoustic materials are employed certain practical and aesthetic considerations should determine their choice:

a: many acoustically absorbent materials are easily damaged and should not be employed within reach,

b: some materials are satisfactory in the first instance but are made ineffective by subsequent decoration,

c: fire resistance has often to be taken into account, and materials which are vermin and rot-proof are obviously desirable,

d: acoustically porous materials may also absorb moisture and the effects of expansion and contraction must be considered,

e: materials with a high coefficient of absorption will obviously be more economical than those which require use in larger quantities. Some of these materials may, however, be found to have a high absorption only over a limited frequency range,

f: some materials may have to be rejected on grounds of

* See also Appendix B.

appearance in the first instance, or because their appearance suffers in the process of fixing or the passage of time.

Since Helmoltz re-discovered the phenomenon of resonant absorption, designers have been able to choose materials of intrinsic beauty and durable qualities, while satisfying the requirements of acoustic absorption. Almost any material may be perforated or slotted so that absorption takes place in a porous material behind. Many such systems are being marketed with published acoustic characteristics.

Finally, the need to 'mix' materials should here be stressed. Since all materials absorb sound preferentially, the aim should be to use them in such proportions that the resultant period of reverberation is about the same at all frequencies.

Acoustics and aesthetics

Acoustics is only one of many aspects in the design of an auditorium and, though acoustics and sight lines are the major considerations, they need not dominate the design.

Over-emphasis of acoustics in the mind of the designer may well lead to loss of character, whether this be the spirit of entertainment in a theatre or the atmosphere of dignity in a council chamber. The occupant of an 'acoustically designed' room may easily feel that he is the subject of a scientific experiment, like an aeroplane in a wind tunnel.

Nevertheless, acoustics should be considered as a positive factor in design, equal with other factors, and not as a matter of 'treatment' when the design is finished.

Only in large auditoria, where projection of sound to distant seats is imperative, need acoustic considerations begin to play a leading role. An 'acoustically shaped' ceiling in a small lecture theatre will be no more effective than a simple horizontal ceiling properly designed. The former may well appear 'out of scale' because the effort made is out of scale with the problem.

Most acoustic problems have alternative solutions, both in general and in detail, so that considerable freedom is left to the designer. For example, dispersive treatment can often take the place of splays if the latter are considered ugly, and reflectors seldom need be terminated exactly at the point determined by geometry if in so doing the result is ill-proportioned.

Acoustics and noise

The noise problem in civic design and building is a considerable field of study in itself, and no attempt is made to deal with it in this book. It cannot of course be divorced from the acoustic design of auditoria, if only because careful acoustic design is vitiated by the interference of intrusive noise.

Reference can only be made here to the need for preventing interference by structure and foundation-borne noise, air-borne noise through windows and doors, the need for sound-absorbent lobbies, acoustic baffles in ventilation trunking and the use of soft flooring materials to reduce audience noise. Otherwise, the reduction of reverberation to what is appropriate for the purpose will do all that is possible in providing a 'quiet room'.

4 : EXAMPLES AND THEIR INDIVIDUAL REQUIREMENTS

Brief mention only is made of the main acoustic considerations in each case, as most have been explained in the previous section, to which reference is made in the margin when applicable.

The examples are in alphabetical order for ease of reference, and each list of requirements is complete in itself.

Bandstands

1: Where these take the form of 'shells' the design is similar to open-air concert platforms, to which reference should be made on page 66. *Refer to page:*

2: Where the audience is seated around a central bandstand, the soffit of the roof forms the main reflector and should be saucer-shaped, presenting its convex surface to the band.

3: Sliding glass screens protect the band from the wind, and act as useful reflectors.

4: The soffit, stage and apron should be designed as resonators. *48*

5: If the bandstand is in a bowl-shaped arena, a paved space around the platform will reflect some sound to the seating area.

6: Tiered seating in a bowl-shaped arena will improve direct sound. *34*

7: If possible, the audience should be protected from wind and external noise by planting.

Churches

Refer to page:

1: The plan and seating arrangement of the church will be much more determined by liturgical considerations than acoustics.

2: Apart from this, acoustical design is made difficult by the conflicting requirements of speech and music, especially in the case of choral and organ music, which require a long period of reverberation. *47*

3: The kind of compromise adopted will depend upon the relative importance of speech and music in each case.

4: Considerable variation in the size of the congregation may be expected in some cases and this will also influence the decision as to a compromise period of reverberation. *47*

5: The height usually required for aesthetic reasons provides a volume per person which makes control of reverberation difficult.

6: An alternative approach to meeting the different requirements of speech and music is to provide a reverberative volume around the choir and organ, while keeping the reverberation as short as possible in the body of the church. The position of the pulpit and lectern in advance of the chancel makes this possible in Anglican churches.

7: If this is done, calculation of reverberation times in the two volumes becomes very approximate, even when the two volumes are clearly defined by a chancel arch. With an open plan, only an average period of reverberation can be calculated as a rough guide.

8: Nevertheless, having in mind the volume per seat, it is unlikely that the error will be on the side of a too low period of reverberation in the body of the church.

9: Where chapels form separate volumes, these may become independent reverberative chambers if their acoustic treatment is not balanced with that of the church. *Refer to page:*

10: Hard, reflective materials around the choir (and organ, if at the chancel end) provide the necessary reverberative conditions.

11: In the body of the church the most convenient position for absorbent materials is likely to be the ceiling or roof lining. Where the section is vaulted or pitched in form, absorbent materials here will reduce the localised reverberation associated with these forms.

12: Apart from this, absorbent materials may with advantage be distributed around wall surfaces, and the wall opposite the pulpit should be absorbent and dispersive. *41*

13: A low angled reflector over the pulpit (the traditional 'sounding board') is essential, and if the pulpit can be placed in a re-entrant angle, so much the better. *39*

Cinemas: normal sound reproduction

1: The plan shape of a cinema is influenced by sight lines to an even greater extent than the theatre, giving usually a longer proportion or a narrower fan shape.

2: The introduction of a balcony tends to reduce the distance from the screen of the furthest seats, where this is required for large audiences. The directional nature of the loudspeaker allows the balcony overhang to be deeper.

3: Although the sound may be raised to any level to reach rear seats, the result may well be unsatisfactory in front seats. Overhead reflectors or the whole ceiling should therefore be designed to provide progressive reinforcement. This can be done more accurately than in the theatre because the position of source is fixed. *Refer to page:* *40*

4: The amplifier should be behind and approximately in the centre of the screen. *41*

5: Side walls should be dispersive, with areas of absorbent material as required to reduce reverberation. *22*

6: The rear wall must be absorbent and dispersive. *43*

7: The rake of the floor may be less than in a theatre, provided clear sight lines are available for all members of the audience, to the bottom of the screen, and provided the amplifier is not too low. *40*

8: Check for echoes from all re-entrant angles. *44*

9: The surfaces of any volume behind the screen should be made absorbent.

10: Allowance should be made for the variable size of audiences, and seats should be as absorbent as possible. *47*

11: Reverberation time should not be more than one second because reverberation is added to the sound track when required. *46*

Cinemas: stereophonic sound

1: The introduction of stereophonic sound changes the whole approach to the acoustic design of the cinema, since amplifiers are distributed all around the auditorium.

2: It is essential that the 'directional effect' is not confused by reflections.

3: This means that all wall surfaces must be absorbent, including the surround of the screen, and preferably also dispersive. *Refer to page:*

4: A 'directional form' for the ceiling is inappropriate and a more or less horizontal and dispersive surface is advisable.

5: Prevention of echoes and standing waves should follow from the above treatment, but the design must be checked for echo-producing corners and reflections from balcony aprons. *44*

6: Whether, in designing a cinema for normal sound reproduction, the possibility of later introduction of stereophonic sound equipment should be taken into account, is a matter for consideration.

7: A cinema designed for stereophonic sound, with a retractable or reversible reflector over the screen, would provide a reasonable compromise. *31*

8: Allowance should be made for the variable size of audiences, and seats should be as absorbent as possible. *47*

9: If possible the cinema should be designed so that amplifiers towards the rear of the auditorium need not be too close to nearby seats.

10: Reverberation time should not be more than one second because reverberation is added to the sound track when required.

Classrooms

1: Though small, a classroom may be much too reverberative if all surfaces are reflective. This is unsuitable for speech and creates a noisy environment.

2: The ceiling, rear wall and the wall opposite the main windows should be made absorbent.

3: Perforated acoustic tiles are suitable for ceilings, soft pinning boards for walls. *Refer to page:*

4: Absorbent material should be introduced to reduce reverberation to $\frac{3}{4}$ second.

Committee rooms

1: Seating arrangements will vary according to the size of the room, but the aim should be to seat members facing each other. Long parallel ranks of seats are therefore undesirable.

2: The volume of the room and the height of the ceiling should not be allowed to become too great in an attempt to create a spacious effect.

3: The volume of the room in relation to the number of persons usually requires the employment of absorbent material to reduce reverberation. It is best distributed around wall surfaces.

4: Where the size of room is sufficient to give rise to near-echoes, dispersive or absorbent treatment of walls and re-entrant ceiling angles may be necessary. *41*

5: A horizontal ceiling provides an equal reinforcement of sound at all positions in the room, and dispersive treatment prevents inter-reflection between committee table and ceiling. *37*

6: Provide well-upholstered seating and soft floor finishes. *47*

7: The reverberation time should be not more than $\frac{3}{4}$ second.

Community-centre halls

1: Ideal acoustics are unattainable in multi-purpose halls for the following reasons:

a: the variety of uses requiring different periods of reverberation, *Refer to page:*

b: the considerable variation in the occupancy of the room,

c: the need to provide a level floor for dancing.

2: Subject to the requirements of sight lines, reduce the length of the room in proportion to its width as much as possible. *33*

3: Provide tiers for the rear seating if possible. *77*

4: A shallow balcony will also provide seating with good acoustics and sight lines, and reduce the number of seats required on the level floor. *77*

5: Employ an angled reflector and the ceiling to reinforce sound at the rear seats, where the angle of incidence of direct sound is small. *39*

6: Provide an apron stage for dramatics and musical entertainment to increase the value of the reflectors.

7: As much resonance as possible should be provided by means of staging, stage apron, resonant reflector and panelling. Resonance compensates to some extent for a short period of reverberation for music. *48*

8: Padded nesting chairs are preferable to canvas or wood seats. *47*

9: If possible arrange for areas to be curtained off at times of small occupancy. *48*

10. Side walls should preferably be dispersive. Window curtains and projecting piers help to meet this requirement, even if only on one side.

11: The rear wall must be absorbent and preferably also *41* dispersive, especially in rooms of long proportion. *Refer to page:*

12: Prevent echoes from the angle between the rear wall and the ceiling. *45*

13: In most cases speech is the more important function and its intelligibility is more critical. A short period of reverberation is also appropriate for films and dances. A reverberation period of about $1\frac{1}{4}$ seconds for the average occupancy expected is therefore a possible compromise, though musical tone will suffer.

Concert halls

1: Various plan shapes can be satisfactory acoustically, but square rather than long proportions are preferable because sound paths are then more equal in length. This helps in obtaining a steady decay of reverberation.

2: A fan-shaped auditorium has the advantage of placing a majority of the audience at some distance from the orchestra, providing them with a more balanced hearing of the orchestral ensemble.

3: If the auditorium is fan-shaped, however, greater care is needed in avoiding a 'megaphone' interior, by the provision of dispersive elements in the design. *30*

4: Seating should be well raked to provide strong direct sound and the clear reception of 'transients'. ('Transient' is the musical term for the initial sound produced by an instrument. It has a different quality from the remainder of a sustained note). *34*

5: Galleries should be shallow to avoid sound shadows and non-reverberative 'pockets'. *45*

6: Provide angled reflectors over, and at the sides of, the platform to provide close reinforcement of direct sound. *31*

7: Reflectors over the platform should be modelled to scatter part of the sound among the players, for mutual appreciation of the orchestral ensemble. *Refer to page:*

8: The ceiling, if shaped, should provide progressive reinforcement of sound towards the rear, but the surface should be broken to contribute towards background reverberation. *39*

9: Side walls should be boldly modelled or broken by galleries or loges. This is especially important if the interior is fan-shaped.

10: The rear wall and gallery aprons should be absorbent and dispersive to prevent echoes. Concave curved walls require a greater degree of absorption. *43*

11: Check carefully for possible echoes from re-entrant angles. *44*

12: As much as possible, absorbent material should be distributed around the interior. This assists in equalising the period of reverberation in different parts of the hall.

13: The over-all aim is to provide strong initial sound by direct path and closely following primary reflection, and a background of steady reverberation.

14: Provide ample resonance in platform staging and panelling. *48*

15: Adjustable absorbent panels may be included to allow for variations in the size of the audience and the type of music. They will also allow for the hall to be 'tuned in' after completion. *48*

16: Absorbent seating and quiet floor finishes are essential. *47*

17: The optimum reverberation varies between $1\frac{1}{2}$ and

$2\frac{1}{4}$ seconds according to the size of the hall.

Refer to page:

Conference halls

1: In the usual arrangement of seating, where the body of the assembly faces the platform, speakers 'from the floor' have their backs turned to part of their audience. A wide rather than deep plan-form reduces the bad effect of this.

2: A curved or splayed arrangement of seating encourages speakers 'from the floor' to turn towards the majority of their audience.

3: Strict economy in seating area and volume is essential if sound amplification is to be avoided in larger halls.

4: A raked floor provides better direct sound paths for speakers on the platform. *34*

5: A low, horizontal reflective ceiling provides reinforcement of sound over the seating area for any position of source. *37*

6: An angled reflector over the platform provides reinforcement of sound at the rear of the hall, and, conversely, reflects sound from the rear of the room to the occupants of the platform. *39*

7: Where galleries are employed, they should be shallow and must not project over the seating area. Occupants of galleries may depend entirely upon ceiling reflections when listening to speakers they cannot see. Seats should therefore not be placed under gallery soffits. *37*

8: All walls should be dispersive or absorbent, preferably both, to avoid echoes or near-echoes relative to the variety of speaking positions. *41*

9: Check carefully for the possibility of echoes from reentrant angles and balcony aprons. *44*

10: Control reverberation by adjustable panels if a considerable variation in the size of conferences is expected. Alternatively, provide for a reduction in the size of the hall for small gatherings. *Refer to page:* 48

11: Provide absorbent seating if possible. 47

12: The continual movement of people, usual at conferences, requires the provision of quiet floor finishes and fixed seating.

13: Reverberation should be in the region of one second, depending on the size of the hall.

Council chambers

1: Subject to the comfortable spacing usually required, form the seating into a compact group.

2: The traditional semi-circular or horse-shoe arrangement of seating is the best acoustically, as the councillors then face each other when speaking.

3: Tiered seating is preferable acoustically and the chairman's table should be on a raised dais. 37

4: Provide a horizontal reflective ceiling over the seating area, as low as reasonably possible. 37

5: The ceiling reflector should be extended to reflect sound into the public gallery, and the gallery should not overhang the main seating area. 37

6: An angled reflector over the dais is optional, but will give the chairman an acoustic advantage in controlling the meeting. 39

7: All walls should be dispersive or absorbent to prevent delayed cross reflections. 41

8: The angles between walls and ceilings should be designed to prevent diagonal echoes. 44

9: The gallery walls and ceiling should be absorbent and quiet floor finishes are essential generally. *Refer to page:*

10: Well upholstered seating reduces the extension of reverberation time at small attendances. *47*

11: The reverberation period should be from $\frac{3}{4}$ to 1 second according to the size of the chamber.

Debating halls

1: Seating arrangements vary considerably according to procedure and the constitution of the assembly, but the over-riding consideration is economy of plan area and volume.

2: Where members face each other, on the House of Commons pattern, rows of seats should be tiered and not too long.

3: The U-shaped arrangement tends to reduce the length of the room for a given number of seats, and has this advantage over the House of Commons arrangement.

4: The ceiling should be horizontal and reflective, and not more than 20 feet above speaking level. It may with advantage be much lower than this in smaller rooms. *37*

5: An angled reflector over the chairman or 'speaker' will assist him in controlling the debate. *39*

6: Where galleries are employed they should be shallow and must not project over the seating area. Occupants of galleries may have to depend upon ceiling reflections when listening to speakers they cannot see. Seats should therefore not be placed under gallery soffits. *37*

7: All walls should be absorbent or dispersive, preferably both, to prevent echoes or near-echoes relative to the variety of speaking positions. *Refer to page:* 41

8: Check carefully for the possibility of diagonal echoes from re-entrant angles. 44

9: Control reverberation by adjustable panels if a considerable variation in the size of meetings is expected. 48

10: Provide absorbent seating and quiet floor finishes. 47

11: Walls and ceilings of galleries should be absorbent to reduce noise.

12: Reverberation should be in the region of 1 second, depending on the size of the hall.

Law courts

1: The customary seating arrangement of law courts takes into account the need to bring the participants as near to each other as possible, for visual and acoustic reasons. This characteristic should therefore be retained.

2: Any difficulty of hearing in law courts is usually the result of excessive reverberation, due to hard reflective surfaces and a large volume per person. 46

3: The ceiling should form a horizontal reflector and be as low as is reasonable in the circumstances. 37

4: An angled reflector over the judge's or magistrates' bench is optional, but will assist in the control of the proceedings. 39

5: The reflective ceiling should be extended towards the public gallery sufficiently to reinforce sound in this area. 37

6: All walls should be absorbent or dispersive to reduce delayed cross reflections. Absorbents used to reduce reverberation should therefore be distributed around the room on wall surfaces above dado height. *Refer to page:* *41*

7: The angles between walls and ceiling should be designed to prevent diagonal echoes. *44*

8: The walls and ceiling of the public gallery should be absorbent and quiet floor finishes should be used.

9: Seating should be upholstered to assist in the maintenance of a low period of reverberation when the number of people present is small. *47*

10: The reverberation period should be not more than 1 second.

Lecture theatres

1: An economic arrangement of seating and gangways is essential. *33*

2: In larger rooms, seating rows splayed around the lecturer's dais will reduce the distance to the back row, but sight lines for the lantern screen will limit the overall width of the seating area.

3: Tiered seating for good acoustics corresponds with good sight lines for the lantern screen and where a demonstration bench is required. *34*

4: A splayed overhead reflector and a horizontal reflective ceiling provides adequate reinforcement of sound in most cases, providing that the ceiling is not too high. *39*

5: If an angled reflector over the dais is not introduced, the ceiling above the lecturer should be dispersive. *48*

6: The walls behind and around the lecturer should act as reflectors when he turns away from his audience.

7: Prevent cross-reflections at the dais by splays or dispersive treatment. *Refer to page:* *42*

8: Side walls should be dispersive or non-parallel. *48*

9: The rear wall should be absorbent and, if curved, also dispersive. *43*

10: Seating or benches should be padded and, if bench fronts are employed, these should be of perforated plywood with absorbent in-filling. *47*

11: Prevent echoes from the angle between rear wall and ceiling. *44*

12: The reverberation period should be in the region of $\frac{3}{4}$ second.

Open-air concert platforms

1: Open-air conditions are not appropriate to orchestral music due to the lack of reverberation. Orchestral music will sound 'thin' as compared with that heard in a concert hall, but, if such an arrangement is required, the following measures should be taken.

2: Compensate as far as possible for lack of reverberation by the use of resonant panels and staging, and by providing the audience with good direct sound and primary reflections. *48*

3: Design the 'orchestra shell' as a megaphone, except that overhead reflectors should be 'broken' so that a small proportion of sound is reflected back to the players.

4: The seating area should be as steeply banked as possible and, if tiered, the rake should be increased for rear seats. *34*

5: The platform should be tiered as in a concert hall.

6: A clear paved space should be allowed in front of the platform but, beyond this, the arrangement of seats and gangways should be economic. *Refer to page:*

7: The Roman theatre arrangement of seating is appropriate, but the plan formation will depend upon the shape of the 'orchestral shell'. A shallow arrangement of reflectors around the orchestra will allow more rows of seats at the sides than if a deeper megaphone shape is employed.

8: Protect the audience from wind and external noise by planting, and in any case choose a quiet site.

Open-air theatres

1: Since reflectors can be used only to a limited extent, dependence on direct sound is greater than in the covered theatre.

2: Seating should be as steeply banked as possible, the rake being increased for the rear rows of seats. *34*

3: Strict economy should be employed in the arrangement of seats and gangways to minimise the distance to the rear seats. *33*

4: No seat should be more than 75 feet from the stage.

5: The classic seating arrangement is appropriate but the number of rows of seats on the transverse axis should be less than on the central axis in the proportion of 3 : 4.

6: An angled reflector above the stage will increase considerably the audibility at rear seats, but this provision will depend upon the character of the theatre. Similarly, any reflecting surfaces around the acting area will be of value, especially when actors are not facing the audience. *39*

7: The traditional paved 'orchestra' space serves as a

reflector if the stage is low and the seating is well raked. Water provides good reflections in this area and precludes subsequent use for additional seating. Alternatively, the 'orchestra' can be used for acting, bringing the actors nearer the audience, and the remainder of the stage kept shallow. *Refer to page:*

8: All acting surfaces should be of wood construction to provide maximum resonance. *48*

9: Protect the audience from wind as much as possible. Wind at 25 m.p.h. reduces by half the distance at which speech can be understood.

10: Protect the theatre from external noise by planting or screen walls, and in any case choose a quiet site.

Opera houses

1: The requirements of the theatre and the concert hall are combined in the design of an opera house and reference should be made under these headings.

2: Italian tradition still influences the shape of the modern opera house and may be said to give it its individual character and a greater 'sense of the occasion'. The fan shape is, however, quite suitable acoustically and makes good sight lines easier to attain.

3: The decision as to shape and gallery arrangement will also be influenced by the type of opera to be performed. The Italian multi-gallery horse-shoe plan provides a high degree of absorption on all wall surfaces and may make the longer period of reverberation required for Wagnerian opera difficult to attain. A reduction in the number and extent of side galleries is therefore indicated. A compromise between the reverberation requirements for Wagnerian and Mozartian operas may be obtained by the provision of projecting and stepped 'boxes' at the sides, with shallow continuous balconies facing the stage.

4: Provided sight lines, and therefore direct sound paths, are good, hearing can be satisfactory even in multi-gallery arrangements, having in mind the somewhat greater strength and clarity of the singing voice, as compared with the conversational speech of the modern theatre. *Refer to page:*

5: Stalls seating should be more steeply raked than is customary, even though this results in a reduction in the number of galleries. *34*

6: Galleries should be kept shallow to avoid sound shadows and non-reverberative pockets. *45*

7: As much assistance as possible should be given to the singers by the provision of overhead reflectors. In multi-gallery arrangements the ceiling may be too high to be effective, in which case it is best made dispersive. A convex reflector, or group of reflectors, over the stage apron should be designed to project reflected sound into all galleries. *39*

8: The provision of an apron stage has two advantages. It makes it possible for the singers to approach nearer the audience, making overhead reflectors more effective and improving direct sound, and it provides a reflector when the singers are 'up-stage'.

9: If the ceiling is high, care should be taken to avoid echoes or near-echoes by the use of dispersive or absorbent treatment. Even shallow domes at high level may cause serious concentrations of delayed reflections. *43*

10: Rear walls, if exposed to direct sound, should be made absorbent. Even at high level, rear walls may cause echoes by secondary reflections between wall and ceiling. *41*

11: Side walls should be dispersive, though the provision of boxes or loges may provide the required modelling of the surface.

12: The orchestra pit may be partly under the apron stage, and should be lined with wood panelling to absorb lower frequencies and add resonance to string instruments. *Refer to page:*

13: Seating should be as absorbent as possible to reduce reverberation at rehearsals. *47*

14: Optimum reverberation varies between $1\frac{1}{4}$ seconds for Mozart and 2 seconds for Wagner, but in most cases a compromise will be necessary. This should favour the requirements for Mozart.

School halls

1: The acoustic design of school halls is made difficult by the same three factors as are mentioned under 'Community Centre Halls', namely:

a: the variety of uses requiring different periods of reverberation,

b: the considerable variation in the occupancy of the room,

c: the need to provide a level floor for dancing, games, parties, etc.

2: Variation in occupancy may be greater than in other multi-purpose halls, and reverberation must be kept down if the room is not to be noisy when used by small groups.

3: A completely plain horizontal and reflective ceiling will give rise to strong inter-reflections between floor and ceiling when the hall is cleared of chairs. After provision has been made for an overhead reflector, the remainder of the ceiling should be absorbent. The reflector itself may with advantage be 'broken'. Angling of the reflector will reduce its size, if kept low over the proscenium. *39*

Refer to page:

4: The average school hall is required to be rectangular to satisfy its various uses, but some larger ones associated with comprehensive schools have been designed on a fan shape. They have the advantage of reducing the distance to rear seats. The notes under the headings 'Theatres' and 'Community Centre Halls' may be more appropriate in such cases.

5: Where the rectangular plan is adopted, side walls should be dispersive and some degree of absorption is usually required to bring reverberation down to a satisfactory level. Structural members and curtains assist towards the above requirement without additional expenditure, but areas of absorbent material may be required on any plain surfaces. *46*

6: The rear wall should be absorbent from ceiling to dado level. *41*

7: If the whole of the floor is not required to be level, a raised or tiered portion at the rear will improve sight lines and acoustics. *77*

8: A shallow gallery will increase the number of seats having better acoustic and visual conditions. When occupied by school children the gallery is inclined to be a source of noise. Walls, floor and ceiling should be absorbent. *77*

9: For dramatic performances, an apron stage has the advantage of bringing young voices out of the absorbent area of scenery and curtains, and underneath the ceiling reflector.

10: Reflectors should be resonant and the stage should be of wood construction with a panelled front. *48*

11: The intelligibility of speech being most critical, reverberation should not be more than 1 second for a full audience, and rather less for small halls. The adopted period of reverberation will also de-

pend upon the calculated results for smaller occupan- *Refer to*
cies. A low period of reverberation will also be suit- *page:*
able for films, games, dances, etc. and only the less frequent musical performances will suffer.

Theatres

1: Seating and gangway arrangements should be as economic as possible, to reduce the distance of rear seats from the stage. *33*

2: Subject to sight lines, a wide, rather than deep auditorium brings the audience nearer the stage. A fan-shaped plan reduces the depth of the auditorium to the minimum for a given number of seats and a given angle of sight line. *33*

3: Balconies also reduce the distance of the furthest seats from the stage, but they must not be so deep as to cause sound shadows over rear seats. *45*

4: For modern acting technique, the furthest seats should not be more than 100 feet from the centre of the acting area. It should also be noted that expressions on actors' faces are indistinguishable beyond 75 feet and this distance is a better acoustic criterion.

5: An apron stage is desirable, not only to encourage acting beneath the overhead reflectors, but also because it acts as a reflector for 'up-stage' positions.

6: 'Open stage' productions provide the best opportunity for good acoustics, but the ceiling or reflector above must be splayed. *39*

7: Stalls seating should be well raked, and at least providing clear sight lines for every member of the audience over the person in front. *34*

8: Balconies should also be raked to provide clear sight lines to the front of the acting area.

9: Overhead reflectors, including the ceiling, should be designed to provide progressively increasing reinforcement of sound towards the rear of the auditorium. Reflectors should be as low as is practical and aesthetically desirable. *Refer to page:* 39

10: For 'proscenium frame' productions, it should be remembered, in designing reflectors, that the 'false proscenium opening' is seldom more than 18 feet high.

11: Side reflectors, unless convex in plan, may cause noticeable fluctuations in the reinforcement of sound as actors move laterally. 38

12: Surfaces not used as reflectors should be dispersive.

13: Rear walls above head level should be absorbent and, if curved, should also be dispersive. 43

14: Balcony aprons, and any other surfaces facing the stage, should be absorbent.

15: Prevent echoes from re-entrant angles at the rear of the auditorium. Such angles can occur in plan as well as section. 44

16: The orchestra pit may be partly under the apron stage, but should be lined with resonant panelling.

17: Seating should be as absorbent as possible. 47

18: Reverberation should be between 1 and $1\frac{1}{2}$ seconds, depending on the size of the theatre, and rehearsal conditions should be taken into account.

Theatre-in-the-round

1: Such theatres are generally smaller than the traditional theatre, and, the audience being on all sides, the number of rows of seats is considerably less.

2: Nevertheless, hearing can be difficult at rear rows when actors have their backs turned. Economy in seating arrangement is therefore still desirable. *Refer to page:*

3: The shape of the auditorium will be largely determined by the shape of the stage and the position of entrances for actors and public. An approximately square (or circular) auditorium equalises the distance to the rear rows of seats, and is better acoustically.

4: Because of the reduced distance to rear seats, reflectors are only required because an actor will always have his back turned to part of the audience. The reflector may take the form of a low horizontal or saucer-shaped 'ceiling' over the stage, or angled reflectors above and on all sides of the stage as shown in fig 26. *37*

5: The remainder of the ceiling should be dispersive.

6: All walls should be absorbent and, if a circular plan is adopted, they should also be dispersive. *41*

7: The stage is generally required to be level with the floor at the first row of seats. To obtain good sight lines, and therefore good acoustics, the seating will have to be more steeply tiered than is usual in the stalls of a traditional theatre. *34*

8: Although one with the floor of the auditorium, the stage must be of resonant wood construction. *48*

9: Seating should be as absorbent as possible. *47*

10: Reverberation should be from $\frac{3}{4}$ to 1 second, dependent upon the size of the auditorium.

Appendix A

EXAMPLE ACOUSTIC ANALYSIS

On the following pages a multi-purpose Civic Hall is taken as an example of the design of an auditorium having acoustic and other considerations in mind. Figs 43 to 46 show the design of the room in plan and section; figs 47 to 49 are a graphic analysis of the reinforcement of sound by primary reflections. There follows a reverberation table and a calculation of the reverberation times for the various functions of the hall.

It is assumed that the hall will be used for dancing, theatrical performances, concerts, cinema shows, conferences, lectures and meetings.

To provide for dancing, part of the floor is level, but this has been kept to the minimum, the remainder being tiered to improve sight lines and acoustics. The hall has been designed with a square proportion to reduce the distance to rear seats and, for the same reason, a gallery has been introduced. Thus, in an auditorium seating 757 persons, no seat is more than 70 feet from the central acting area. The side galleries would be used only for conferences, meetings and apron stage productions.

The stage also is designed with multi-purpose use in mind, the apron stage having four positions and being operated by hydraulic lift (fig 44). In the lower position it provides for an orchestra pit; level with the floor it provides space for additional seating (shown dotted on fig 43); at an intermediate position it provides the first tier for an orchestra or dance band; level with the stage it provides an apron or proscenium.

The arrangement for dancing is shown on fig 45, the sprung floor being placed laterally to the band platform, and the surrounding tiers being used for sitting out. Tables are hinged and recessed into the backs of the solid balustrades when not required.

Acoustically, the room is designed to have reverberation periods suitable for conferences, drama, cinema and dances. For concerts the reverberation time would be short and, for this reason, the walls are lined with wood panelling to add to the resonance of stage and proscenium reflector.

The ceiling and reflector are designed to provide progressively increasing reinforcement of sound towards the rear seats (fig 47), and the rear splay is returned along the flank walls to provide reinforcement of sound in the side galleries (figs 48 and 49). Galleries project only so far as will avoid sound shadows being formed thereunder. Re-entrant angles at the rear of the auditorium have absorbent or dispersive treatment on at least one surface, and the glazed screen at the entrance to the room is angled on plan to disperse echoes.

The reverberation table provides a brief specification of the materials employed and reverberation is calculated for high, low and middle frequencies.

A Civic Hall

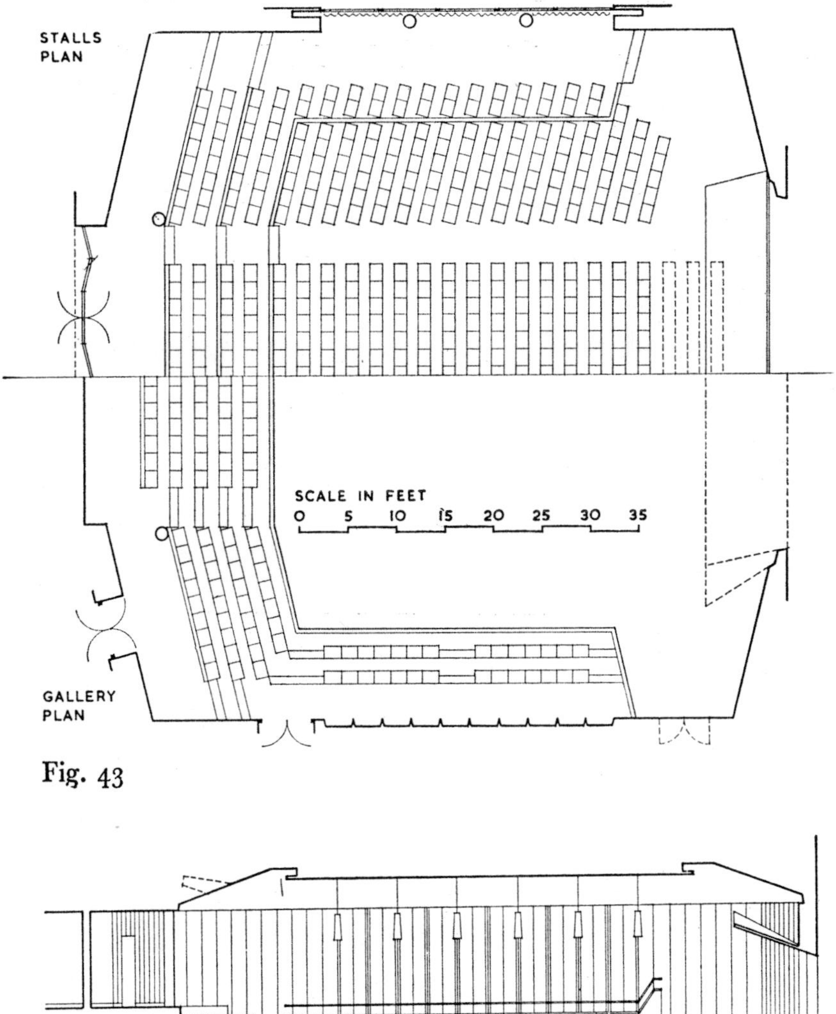

Fig. 43

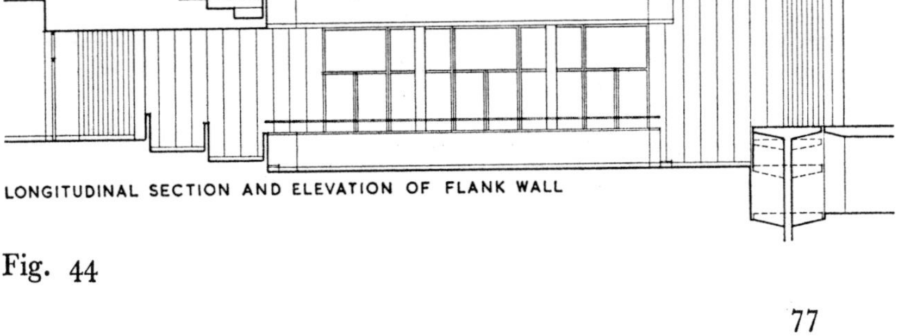

Fig. 44

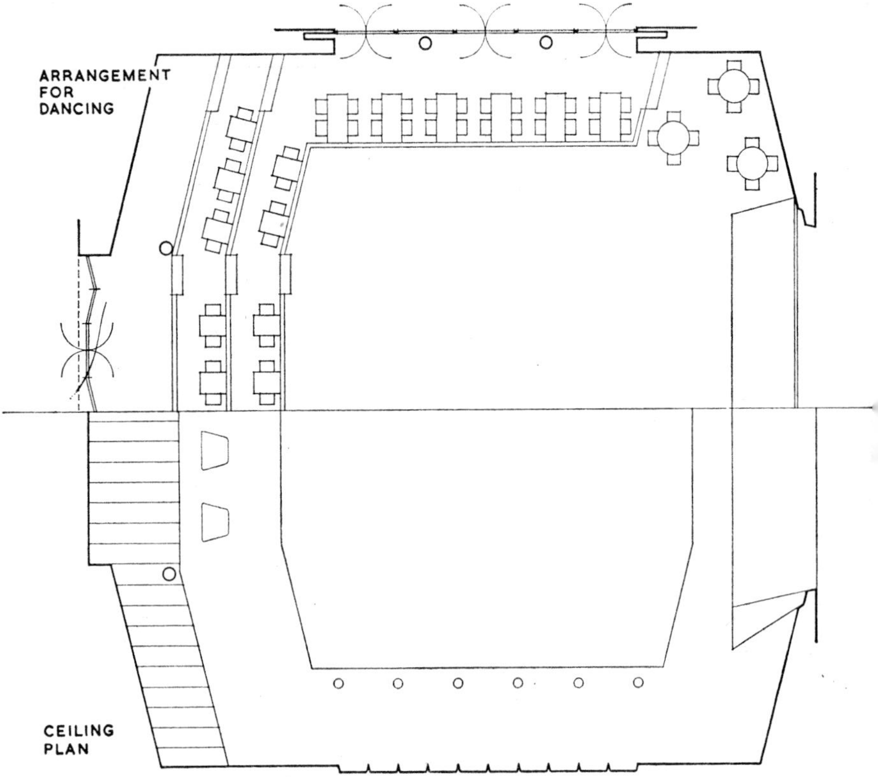

Fig. 45

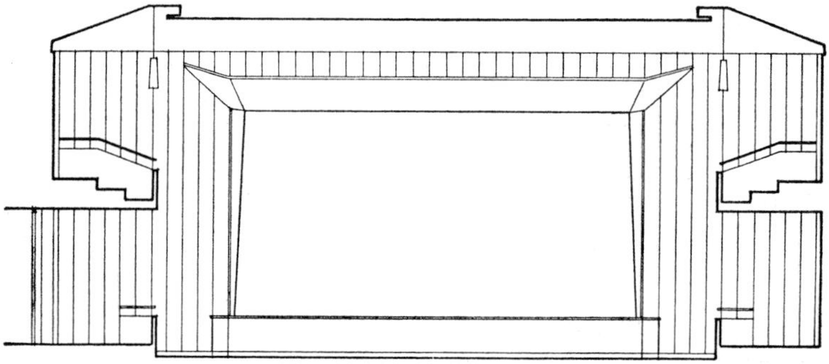

CROSS SECTION AND ELEVATION OF PROSCENIUM WALL

Fig. 46

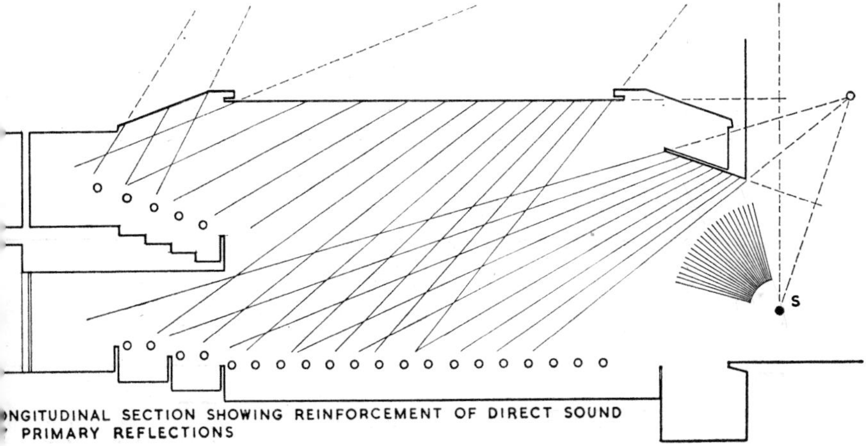

NGITUDINAL SECTION SHOWING REINFORCEMENT OF DIRECT SOUND
PRIMARY REFLECTIONS

g. 47

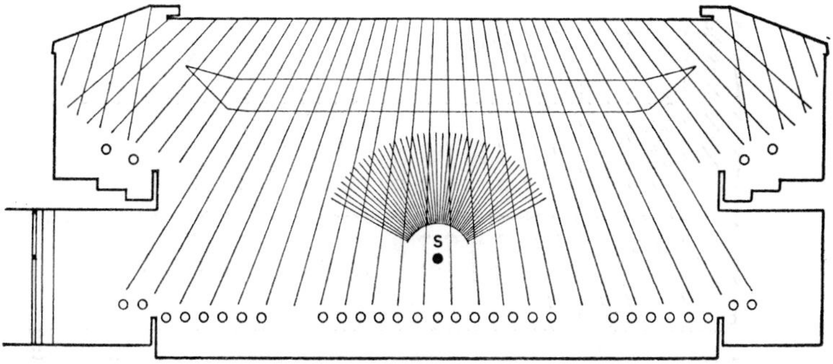

CROSS SECTION SHOWING CEILING REFLECTIONS

Fig. 48

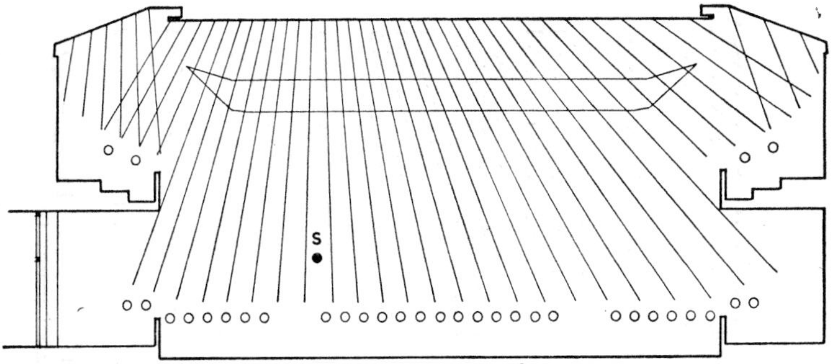

CROSS SECTION. SOUND SOURCE AT SIDE OF STAGE

Fig. 49

Reverberation Calculation · Civic Hall

VOLUME: 123000 cu. ft. MAXIMUM AUDIENCE: 757.

ITEM, POSITION AND DESCRIPTION.	VOLUME AREA OR NUMBER	COEFF. 125 C.	O.W.U.	COEFF. 500 C.	O.W.U.	COEFF. 2000 C.	O.W.U.
Air. (negligible absorption below 1000 c.).	123000	—	—	—	—	0·002	246
Central suspended ceiling. Plaster on metal lath. Air space	2000	0·20	400	0·10	200	0·04	80
Splayed " " " " " " " "	2700	0·20	540	0·10	270	0·04	108
Balcony soffit. Plaster on solid backing.	1825	0·03	54·75	0·02	36·5	0·04	73
Rear ceiling over balcony. Plaster. Dispersive.	530	0·03	15·9	0·02	10·6	0·04	21·2
Splayed reflector over proscenium. Veneered blockboard.	370	0·03	111	0·10	37	0·10	37
Level floor. Maple wood strips on battens. Sprung.	2350	0·15	352·5	0·10	235	0·08	188
Tiered floor. Cork tiles on concrete and screed. Shading reduction.	1825	0·04	73	0·05	91·25	0·02	36·5
Balcony tiers. Cork tiles on concrete and screed. Shading reduction.	1825	0·04	73	0·05	91·25	0·02	36·5
Proscenium stage. Oak strips on staging.	350	0·15	52·5	0·10	35	0·08	28
Proscenium opening. Curtains or scenery.	630	0·20	126	0·30	189	0·40	252
Walls at side of proscenium opening. Veneered plywood panelling.	880	0·25	220	0·15	132	0·10	88
Walls flanking proscenium opening. Veneered plywood panelling.	625	0·25	156·25	0·15	93·75	0·10	62·5
Dispersive panels at gallery level. Veneered plywood panelling.	660	0·25	165	0·15	99	0·10	66
Rear side walls. Veneered plywood panelling.	710	0·25	177·5	0·15	106·5	0·10	71
Dispersive panel below side gallery. Veneered plywood panelling.	350	0·25	87·5	0·15	42·5	0·10	35
Screen below side gallery. Plate glass in metal frame.	350	0·10	35	0·04	14	0·02	7
Rear wall at gallery level. Perforated panelling. Absorbent quilt in air space.	315	0·35	110·25	0·80	252	0·65	204·75
Entrance screen and doors. Plate glass in metal frame.	300	0·10	30	0·04	12	0·02	6
Rear walls at side of entrance. Perforated panelling. [illegible]	360	0·35	126	0·80	288	0·65	234

in air space.							
Balustrades at lower level. Hardwood strips on framing.	395	0·25	98·75	0·50	197·5	0·45	177·75
	490	0·15	73·5	0·10	49	0·05	24·5
Ventilation grilles. Metal. 50% voids.	100	0·15	15	0·35	35	0·30	30
PERMANENT ABSORPTION:			3093·40		2516·85		2112·45
VARIABLE FACTORS:							
Curtains over glazed screen. Medium weight. Additional coeff.	350	0·05	17·5	0·31	108·5	0·48	168
Seating.							
Fabric upholstered nesting chairs at lower level.	568	2·00	1136	3·00	1704	3·00	1704
Fabric upholstered theatre seats in gallery.	189	2·00	378	3·00	567	3·00	567
Audience.							
Maximum for Conferences. Coeff. additional to seating.	757	1·00	757	2·00	1514	1·50	1135·5
Drama or Cinema. Side gallery seats vacant.	701	1·00	701	2·00	1402	1·50	1051·5
Dancing. Dancers, spectators and band.	310	3·00	930	5·00	1550	4·50	1395
Rehearsals. Say 10 persons. Additional coeff.	10	1·00	10	2·00	20	1·50	15
TOTAL ABSORPTION FOR: (by adding in variable factors as appropriate)							
CONFERENCES			5364·4		6301·85		5518·95
DRAMA AND CINEMA			5325·9		6298·35		5602·95
DANCES			4311·4		4633·85		4074·45
REHEARSALS			4617·4		4807·85		4398·45
REVERBERATION TIME IN SECONDS FOR:							
CONFERENCES		1·15		0·98		1·11	
DRAMA AND CINEMA		1·15		0·98		1·10	
DANCES		1·43		1·33		1·51	
REHEARSALS		1·33		1·28		1·40	
(by formula : $t=\frac{V}{A}\times 0{\cdot}05$)							
VOLUME PER PERSON: 162 cu. ft.							

Appendix B

COEFFICIENTS OF ABSORPTION

In the following schedule composite materials are alphabetically classified according to their *surface* finishes. Special acoustic materials are marked by an asterisk and many of these are proprietary.

Coefficients are given for 1 square foot of the material unless otherwise stated.

		125^c	500^c	2000^c
AIR	per cubic foot:	—	—	0·003
*ASBESTOS	½″ sprayed on solid backing:	0·10	0·40	0·60
*ASBESTOS	1″ sprayed on solid backing:	0·15	0·50	0·70
*ASBESTOS FELT	1″ muslin covered, one coat distemper:	0·11	0·69	0·59
*ASBESTOS FELT	1″ muslin covered, painted & perforated:	0·10	0·77	0·60
*ASBESTOS TILES	3/16″ perforated asbestos cement on battens, 1″ mineral wool in air space:	0·17	0·94	0·65
*ASBESTOS TILES	3/16″ perforated asbestos cement on battens, 2″ mineral wool in air space:	0·29	0·94	0·67
*ASBESTOS TILES	⅛″ perforated asbestos cement backed with 1″ rock wool pad, on battens, 1″ free air space:	0·32	0·72	0·76
AUDIENCE	per person, in upholstered theatre seats:	3·0	5·0	4·0
AUDIENCE	per person, in wood chairs:	2·0	4·0	4·5
BRICKWORK	open texture, unpainted:	0·02	0·03	0·04
BRICKWORK	dense texture, unpainted:	0·01	0·02	0·03
BRICKWORK	painted two coats:	0·01	0·02	0·03

		125c	500c	2000c
CANVAS	stretched, medium weight, 6″ air space:	0·10	0·25	0·30
CANVAS	in oil paintings, unglazed:	0·10	0·28	0·30
CARPET	thin, on thin felt, on solid floor:	0·10	0·25	0·30
CARPET	thick, on thick felt, on solid floor:	0·10	0·30	0·50
CARPET	thick, on boards and joists or battens, free air space:	0·20	0·30	0·50
CHAIRS	see under 'Seating'			
CLINKER BLOCKS	unplastered:	0·20	0·60	0·50
CONCRETE	rough finish:	0·02	0·02	0·05
CONCRETE	smooth finish:	0·01	0·02	0·02
CORK TILES	¾″ bedded solid, wax polished:	0·05	0·07	0·09
CURTAINS	in folds, against wall, light weight:	0·05	0·10	0·20
CURTAINS	in folds, against wall, medium weight:	0·07	0·35	0·50
CURTAINS	in folds against wall, heavy weight:	0·10	0·50	0·60
CURTAINS	dividing spaces, medium weight:	0·03	0·10	0·20
FIBREBOARD	½″ wood or cane fibre, soft texture, solid mounting, undecorated:	0·05	0·15	0·30
FIBREBOARD	½″ wood or cane fibre, soft texture, solid mounting, two coats distemper:	0·05	0·10	0·10
FIBREBOARD	½″ wood or cane fibre, soft texture, on battens, 1″ free air space, undecorated:	0·30	0·35	0·30
FIBREBOARD	½″ wood or cane fibre, soft texture, on battens, 1″ free air space, two coats distemper:	0·30	0·15	0·10
*FIBREBOARD TILES	½″ perforated, cane fibre, on battens, 1″ free air space:	0·10	0·45	0·40

		125c	500c	2000c
*FIBREBOARD TILES	¾″ perforated, cane fibre, on battens, 1″ free air space:	0·20	0·65	0·60
*FIBREBOARD TILES	1¼″ perforated, cane fibre, on battens, 1″ free air space:	0·20	0·85	0·55
*FIBREBOARD TILES	½″ perforated, cane fibre, solid mounting:	0·10	0·40	0·45
*FIBREBOARD TILES	¾″ perforated, cane fibre, solid mounting:	0·10	0·70	0·65
*FIBREBOARD TILES	1¼″ perforated, cane fibre, solid mounting:	0·10	0·85	0·65
*FIBREBOARD TILES	½″ perforated, wood fibre, on battens, 1″ free air space:	0·20	0·60	0·65
*FIBREBOARD TILES	⅝″ perforated, wood fibre, on battens, 1″ free air space:	0·25	0·60	0·70
*FIBREBOARD TILES	¾″ perforated, wood fibre, on battens, 1″ free air space:	0·25	0·65	0·80
*FIBREBOARD TILES	slotted softboard on channelled softboard, total 1″ thick:	0·05	0·70	0·95
*FIBREBOARD TILES	½″ perforated wood fibre, decorated, on battens, 1″ free air space:	0·09	0·57	0·42
*FIBREBOARD TILES	¾″ perforated wood fibreboard over channelled wood fibre board, on battens, 1″ free air space:	0·21	0·66	0·80
*FIBREBOARD TILES	¾″ channelled wood fibreboard, on battens, 1″ free air space:	0·15	0·58	0·82
*FIBREBOARD TILES	1″ channelled wood fibreboard, on battens, 1″ free air space:	0·35	0·50	0·75
*FIBREBOARD TILES	¾″ cross-channelled fibreboard, on battens, 1″ free air space:	0·20	0·65	0·88

		125c	500c	2000c
GLASS	in windows, up to 32 oz.:	0·30	0·10	0·07
GLASS	in windows, $\frac{1}{4}''$ plate:	0·10	0·05	0·02
GLASS	bedded solid:	0·01	0·01	0·02
*GLASS FIBRE	1″ resin bonded:	0·10	0·55	0·75
*GLASS FIBRE	2″ resin bonded:	0·20	0·70	0·75
GRANOLITHIC	on solid floor:	0·01	0·02	0·02
HARDBOARD PANELLING	on battens, 1″ free air space:	0·20	0·15	0·10
*HARDBOARD PANELLING	10% perforated, 1″ fibre-glass in air space:	0·10	0·55	0·80
*HARDBOARD PANELLING	10% perforated, 2″ fibre-glass in air space:	0·20	0·75	0·70
*HARDBOARD PANELLING	5% perforated, 1″ fibre-glass in air space:	0·10	0·75	0·30
*HARDBOARD PANELLING	5% perforated, 2″ fibre-glass in air space:	0·20	0·95	0·25
*HARDBOARD TILES	perforated, on $\frac{3}{4}''$ channelled wood fibre board, on battens, 1″ free air space:	0·21	0·66	0·80
*HARDBOARD TILES	perforated, on $\frac{3}{4}''$ perforated wood fibre board, on battens, 1″ free air space:	0·30	0·40	0·50
*HARDBOARD TILES	perforated, on $\frac{3}{4}''$ wood frame, mineral wool infilling, mounted over air space:	—	0·77	—
LINOLEUM	on solid floor:	0·05	0·05	0·10
MARBLE	on solid backing:	0·01	0·01	0·02
*METAL PANELLING	3% perforated, 1″ fibre-glass in air space:	0·10	0·70	0·40
*METAL PANELLING	3% perforated, 2″ fibre-glass in air space:	0·20	0·90	0·25
*METAL PANELLING	10% perforated, 1″ fibre-glass in air space:	0·10	0·55	0·80
*METAL PANELLING	10% perforated, 2″ fibre-glass in air space:	0·20	0·75	0·70
*METAL TILES	$1\frac{1}{4}''$ perforated metal trays, mineral wool pads, 2″ free air space:	0·20	0·80	0·80
*METAL TILES	perforated stove-enamel-			

		125^{c}	500^{c}	2000^{c}
	led metal tray, 2½″ mineral wool pad, free air space:	0·25	0·96	0·85
OPENINGS, GRILLES	ventilation grilles, 50% voids:	0·15	0·35	0·30
OPENINGS, STAGE	proscenium opening, average curtains or scenery:	0·20	0·30	0·40
OPEN WINDOWS	per square foot:	1·00	1·00	1·00
PLASTER	lime or gypsum, on solid backing:	0·02	0·02	0·04
PLASTER	lime or gypsum, on lath, small air space:	0·03	0·01	0·04
PLASTER	suspended, on metal lath, large air space:	0·02	0·10	0·04
*PLASTER	acoustic, ½″ on solid backing:	0·13	0·35	0·45
PLASTER-BOARD	on firring, small air space:	0·03	0·01	0·04
*PLASTER-BOARD	12% perforated, 1″ mineral wool and 1″ free air space:	0·17	0·90	0·45
*PLASTER-BOARD	12% perforated, 2″ mineral wool and 1½″ free air space:	0·37	0·85	0·45
*PLASTER TILES	¾″ perforated fibrous plaster, aluminium foil backing, free air space:	0·45	0·80	0·65
*PLASTER TILES	⅝″ perforated gypsum, fibreglass tissue backing, 1″ air space:	0·20	0·65	0·35
*PLASTER TILES	⅝″ perforated gypsum, 1″ fibreglass quilt, 1″ air space:	0·30	0·80	0·45
*PLASTER TILES	skim coat plaster on 1″ perforated asbestos fibre tile, 1″ air space:	—	0·75	—
*PLASTER TILES	1″ acoustic plaster, on solid backing:	0·15	0·40	0·50
*PLASTIC CLOTH	perforated plastic cloth on ¾″ perforated wood fibre-			

		125c	500c	2000c
	board, on battens, 1″ free air space:	0·30	0·70	0·75
*PLASTIC CLOTH	perforated plastic cloth on 1″ asbestos felt, perforations $\frac{3}{32}$″ diam. 10 per sq. in.:	0·11	0·67	0·58
*PLASTIC TILES	$\frac{1}{2}$″ expanded poly-styrene, unperforated, 1″ air space:	0·05	0·40	0·20
*PLASTIC TILES	$\frac{3}{4}$″ expanded poly-styrene, perforated, 1″ air space:	0·05	0·70	0·20
RUBBER	flooring, on solid:	0·05	0·05	0·10
SEATING	wood chairs, per chair:	0·15	0·17	0·20
SEATING	metal and canvas chairs, per chair:	0·80	1·60	1·90
SEATING	theatre seats, leather padded, per seat:	1·20	1·60	1·90
SEATING	theatre seats, fabric upholstered back and seat, perforated under-seat, (Festival Hall type): per seat:	2·00	3·00	3·00
SEATING	theatre seats, leather or plastic upholstered back and seat, perforated underseat, per seat:	2·00	2·50	2·50
SEATING	council chamber seats, heavily upholstered, fabric-covered, per seat:	2·80	3·00	3·30
SHADING	reduction of coefficient for floor finish where floor is 'shaded' by seating:	20%	40%	60%
STAGING	see under 'wood staging'			
STEEL	see under 'metal'			
STONE	natural finish:	0·01	0·02	0·02
STONE	polished:	0·01	0·01	0·02
TAPESTRY	stretched over $1\frac{1}{2}$″ air space:	0·05	0·25	0·30
TERRAZZO	on walls or floors:	0·01	0·01	0·02
TILES	glazed:	0·01	0·01	0·02
TILES	unglazed:	0·03	0·03	0·05

		125c	500c	2000c
WATER		0·01	0·01	0·02
WINDOWS	see under 'glass' and 'openings'			
WOOD BLOCKS	on solid floor:	0·02	0·05	0·10
WOOD BOARDING	$\frac{3}{4}$″ match boarding on battens on solid wall:	0·30	0·10	0·10
WOOD FLOORING	1″ boards on joists or battens, free air space:	0·10	0·10	0·10
WOOD PANELLING	3 ply on solid backing:	0·10	0·10	0·10
WOOD PANELLING	3 ply on battens, 1″ free air space:	0·30	0·15	0·10
*WOOD PANELLING	3 ply on battens, 1″ acoustic felt in air space:	0·40	0·15	0·10
*WOOD PANELLING	3 ply on battens, 2″ acoustic felt in air space:	0·50	0·25	0·20
*WOOD PANELLING	15% perforated, 3 ply, on battens, 2″ fibreglass in air space:	0·20	0·60	0·35
*WOOD PANELLING	15% perforated, 3 ply, 2″ fibreglass and 1″ free air space:	0·33	0·80	0·40
*WOOD PANELLING	5% perforated, 3 ply, 2″ fibreglass and 1″ air space:	0·40	0·63	0·13
WOOD STAGING	1″ boards on staging, large air space:	0·15	0·10	0·10
*WOOD STRIPS	1$\frac{1}{2}$″ wood strips, spaced, $\frac{3}{16}$″ slots, 1″ mineral wool in air space:	0.25	0.65	0.65
*WOOD VENEERED TILES	perforated, on perforated fibre board, on battens, 1″ free air space:	0·20	0·50	0·75
WOODWOOL SLABS	1″ unplastered on solid backing:	0·10	0·40	0·60
WOODWOOL SLABS	3″ unplastered on solid backing:	0·20	0·80	0·80
WOODWOOL SLABS	1″ unplastered, on battens, on solid wall, 1″ free air space:	0·10	0·60	0·60

INDEX

Materials are not individually indexed. For an alphabetical list of acoustic materials, see Appendix B. The figures in the index refer to page numbers.